AF839703

Enhanced Characterization and Representation of Flow Through Karst Aquifers

Awwa Research Foundation

About the Awwa Research Foundation

The Awwa Research Foundation (AwwaRF) is a member-supported, international, nonprofit organization that sponsors research to enable water utilities, public health agencies, and other professionals to provide safe and affordable drinking water to consumers.

The Foundation's mission is to advance the science of water to improve the quality of life. To achieve this mission, the Foundation sponsors studies on all aspects of drinking water, including supply and resources, treatment, monitoring and analysis, distribution, management, and health effects. Funding for research is provided primarily by subscription payments from approximately 1,000 utilities, consulting firms, and manufacturers in North America and abroad. Additional funding comes from collaborative partnerships with other national and international organizations, allowing for resources to be leveraged, expertise to be shared, and broad-based knowledge to be developed and disseminated. Government funding serves as a third source of research dollars.

From its headquarters in Denver, Colorado, the Foundation's staff directs and supports the efforts of more than 800 volunteers who serve on the board of trustees and various committees. These volunteers represent many facets of the water industry, and contribute their expertise to select and monitor research studies that benefit the entire drinking water community.

The results of research are disseminated through a number of channels, including reports, the Web site, conferences, and periodicals.

For subscribers, the Foundation serves as a cooperative program in which water suppliers unite to pool their resources. By applying Foundation research findings, these water suppliers can save substantial costs and stay on the leading edge of drinking water science and technology. Since its inception, AwwaRF has supplied the water community with more than $300 million in applied research.

More information about the Foundation and how to become a subscriber is available on the Web at **www.awwarf.org.**

Enhanced Characterization and Representation of Flow Through Karst Aquifers

Prepared by:
Scott L. Painter, **Alexander Sun**, and **Ronald T. Green**
CNWRA
Southwest Research Institute®
6220 Culebra Road, San Antonio, Texas 78238-5166

Jointly sponsored by:
Awwa Research Foundation
6666 West Quincy Avenue, Denver, CO 80235-3098

Southwest Florida Water Management District

and

Edwards Aquifer Authority

Published by:

Awwa Research Foundation

American Water Works Association

IWA Publishing

DISCLAIMER

This study was jointly funded by the Awwa Research Foundation (AwwaRF), the Southwest Florida Water Management District (SFWMD), and the Edwards Aquifer Authority (EAA). AwwaRF, SFWMD, and EAA assume no responsibility for the content of the research study reported in this publication or for the opinions or statements of fact expressed in the report. The mention of trade names for commercial products does not represent or imply the approval or endorsement of AwwaRF, SFWMD, or EAA. This report was presented solely for informational purposes.

CONTENTS

TABLES

FIGURES

FOREWORD

The Awwa Research Foundation is a nonprofit organization that is dedicated to the implementation of a research effort to help utilities respond to regulatory requirements and traditional high-priority concerns of the industry. The research agenda is developed through a process of consultation with subscribers and drinking water professionals. Under the umbrella of a Strategic Research Plan, the Research Advisory Council prioritizes the suggested projects based upon current and future needs, applicability, and past work; the recommendations are forwarded to the Board of Trustees for final selection. The foundation also sponsors research projects through the unsolicited proposal process; the Collaborative Research, Research Applications, and Tailored Collaboration programs; and various joint research efforts with organizations such as the U.S. Environmental Protection Agency, the U.S. Bureau of Reclamation, and the Association of California Water Agencies.

This publication is a result of one of these sponsored studies, and it is hoped that its findings will be applied in communities throughout the world. The following report serves not only as a means of communicating the results of the water industry's centralized research program but also as a tool to enlist the further support of the nonmember utilities and individuals.

Projects are managed closely from their inception to the final report by the foundation's staff and large cadre of volunteers who willingly contribute their time and expertise. The foundation serves a planning and management function and awards contracts to other institutions such as water utilities, universities, and engineering firms. The funding for this research effort comes primarily from the Subscription Program, through which water utilities subscribe to the research program and make an annual payment proportionate to the volume of water they deliver and consultants and manufacturers subscribe based on their annual billings. The program offers a cost-effective and fair method for funding research in the public interest.

A broad spectrum of water supply issues is addressed by the foundation's research agenda: resources, treatment and operations, distribution and storage, water quality and analysis, toxicology, economics, and management. The ultimate purpose of the coordinated effort is to assist water suppliers to provide the highest possible quality of water economically and reliably. The true benefits are realized when the results are implemented at the utility level. The foundation's trustees are pleased to offer this publication as a contribution toward that end.

Walter J. Bishop
Chair, Board of Trustees
Awwa Research Foundation

Robert C. Renner, P.E.
Executive Director
Awwa Research Foundation

ACKNOWLEDGMENTS

The authors of this report thank the Awwa Research Foundation, Edwards Aquifer Authority, Southwest Florida Water Management District, Miami-Dade Water and Sewer Department, and San Antonio Water System for supporting this work. The authors acknowledge James Winterle, Wesley Patrick and David Ferrill for careful reviews of the report. The authors are also grateful to Cheryl Patton for formatting the report.

This report is an independent product of the Center for Nuclear Waste Regulatory Analyses (CNWRA) and does not necessarily reflect the views or positions of the funding organizations.

QUALITY ASSURANCE STATEMENT

No CNWRA-generated original data are contained in this report. The analyses contained in this report are documented according to CNWRA quality assurance procedures in Scientific Notebooks SN660E and SN634E.

EXECUTIVE SUMMARY

Numerical flow models are important tools for effective management of groundwater resources. Representing groundwater flow through karst aquifers is challenging because of the highly heterogeneous nature of groundwater flow through a medium with multiple interacting flow systems (i.e., flow through the matrix, fractures, and conduits). This jointly sponsored karst modeling project was organized to develop enhanced groundwater modeling capabilities that can be used to effectively manage karst aquifers. This report summarizes the first year of the multi-year project.

The first major tasks in this project were to identify key considerations for modeling karst aquifers, evaluate existing modeling approaches, and select an approach for detailed investigation and development during the remainder of the project. Three key considerations for modeling karst systems were identified: representation of conduits in flow models, representation of flow dynamics in conduits, and representation of partially filled conduits.

After surveying the existing distributed parameter models, three broad approaches for representing conduits were identified: embedded discrete-conduit models; single-continuum, smeared-conduit models; and dual-conductivity models. Embedded discrete-conduit models are literal representations of pipe-like conduits embedded in a diffuse flow system and are intrinsically linked to the finite-element numerical model. Single-continuum, smeared-conduit models attempt to capture the effects of a conduit on the groundwater flow by assigning high transmissivity and low storage to grid cells crossed by a conduit. This approach is how karst conduit flow is usually represented in the MODFLOW (Harbaugh et al., 2000) groundwater modeling software. In dual-conductivity approaches, two interacting flow systems representing conduits and a background diffuse flow system are modeled.

A series of numerical experiments was used to identify the limitations and advantages of the three general representations of conduits within groundwater models. The commercial code FEFLOW® (Diersch, 2002) was used for the embedded discrete-conduit approach. MODFLOW-2000 was used for the single-continuum, smeared-conduit simulations. A new dual-conductivity MODFLOW package (DCM) was developed for testing the dual-conductivity approach. Results of the numerical experiments show that the smeared-conduit approaches are surprisingly flexible and can mimic the discrete-conduit models for a wide range of conditions. However, the single-continuum, smeared-conduit approach becomes inaccurate if the density of conduits is large. In addition, MODFLOW-2000 does not include the capability for modeling turbulent flow in open conduits, which is considered necessary for adequate representation of flow dynamics within karst conduits. The DCM package provides more modeling flexibility compared with the single-continuum, discrete-conduit approach and was able to reproduce the discrete-conduit results in all three numerical experiments. In addition, the DCM model has an additional calibration parameter, the conduit-background exchange term, which is not needed in the other two approaches.

Although intuitively appealing, the literal representation of embedded discrete conduits within a finite-element code is not recommended because existing finite-element representations (e.g., using FEFLOW®) are not adequate for unconfined karst aquifers. In addition, finite-element codes are not compatible with legacy groundwater modeling data, which are typically in the MODFLOW format. Also, field methods for determining positions and detailed geometry of conduits are not available.

An interim project meeting was held on April 5, 2004, at the Marshall Center at the University of South Florida. There were 24 participants at the meeting including the core project team, representatives from the participating institutions, and representatives from the funding agencies. The primary objective of that meeting was to select a modeling approach for development during the remainder of this project. The consensus among the attendees was that the DCM package should be developed further and released to the project team for use in the second- and third-year modeling activities.

Consequently, the second half of the first year of this project was devoted to further developing, documenting, and testing the dual-conductivity DCM package for use with MODFLOW-2000. A turbulence model was added, the software was validated using simple benchmark problems, and code documentation was completed. In addition, demonstration simulations of the Barton Springs Segment of the Edwards Aquifer near Austin, Texas, were undertaken. The results of those simulations demonstrate that the modeling tool can be successfully applied in complex applications.

CHAPTER 1
INTRODUCTION

BACKGROUND

The Awwa Research Foundation (AwwaRF)-sponsored karst modeling project was organized to develop tools for enhanced characterization and representation of flow through karst aquifers. The project is referred to hereafter as the karst modeling project. The primary motivation of the project is to provide tools for effectively managing karst aquifers. Management tools for karst aquifers should be able to help in the

- Development of guidelines for aquifer-scale groundwater flow models that properly incorporate fast groundwater velocities associated with karst conduit flow
- Evaluation of groundwater tracer test results
- Identification of realistic wellhead and springhead protection areas in karst aquifers
- Assessment of contaminant spills over karst aquifers
- Evaluation of the efficacy of recharge credits
- Evaluation of recharge and recirculation schemes to augment spring discharge rates
- Evaluation of boundary conditions for karst aquifers, including inland surface water features and tidal effects

Aquifer management is best performed using numerical models that provide representative groundwater flow simulation. Difficulties can arise when water resource managers attempt to replicate conduit flow through karstified media using flow simulators based on Darcy's law. Although Darcy's law-based simulators can potentially simulate groundwater flow through a karst system for some conditions, typical application of Darcy's law-based simulators, such as the standard version of MODFLOW (Harbaugh et al., 2000), may have difficulty capturing flow transients in karst aquifers. This limitation can potentially result in miscalculations when attempting to achieve the karst aquifer management goals listed above. Advances made by this project are intended to provide water resource managers with the tools to correctly simulate groundwater flow through karst aquifers to ensure water resource management goals are properly met.

OBJECTIVES

The long-term (3- to 4-year) objective of this project is to develop new modeling approaches and tools to address applications involving karst aquifers with significant conduit flow. The objective of the first year is to develop quantitative tools for modeling groundwater flow in aquifers with explicit conduits.

SCOPE

The original work plan for the multiyear project calls for major tasks addressing model evaluation, model development, and applications. Model evaluation includes becoming familiar with the functionality, operation, capabilities, and limitations of existing modeling approaches and computer codes, and identifying a path forward for developing new capabilities. Model development has six main components: (i) conceptualization, (ii) prototyping, (iii) implementation, (iv) documentation, (v) testing, and (vi) demonstration. Not all of these components are within the scope of the first year of this project. Model conceptualization includes identifying a modeling approach, mathematical equations, and computer algorithms. Prototyping involves preliminary implementation for the purpose of algorithm development and testing. Implementation refers to translating the mathematical model to computer software and may include modifying existing software. Testing involves applying the model to a synthetic data case. Demonstration entails using the model to represent a relatively well-defined actual, albeit simple, physical system. Applications to full-scale physical systems with complicated, poorly or underdefined physical characteristics are planned. This approach is to provide a logical path from the inception of model identification to the successful replication of a karst aquifer.

This report addresses model evaluation, conceptualization, prototyping, implementation, documentation, and testing. Preliminary model demonstration results are also presented. Model demonstration and applications will be continued in the later years of the project.

This report is limited to distributed-parameter groundwater models. Lumped parameter models in which an aquifer is modeled as a small number of coupled storage volumes are not considered, although these may have applications in predicting spring flows.

PROJECT ACCOMPLISHMENTS

Analyses conducted soon after the start of this project resulted in a minor adjustment to the first-year project plan, but not to the long-term project objective. Early results suggested that conduit flow dynamics could be captured with modifications to existing modeling approaches. This adjustment to the first-year project plan did not affect the objectives, but extended the scope of potential conduit flow modeling tools to include volume-averaged or smeared-modeling approaches in addition to finite-element conduit models. Modeling approaches evaluated during this first task were based on adaptations of the MODFLOW code and a recent release (May 2002) of the finite element simulator FEFLOW® (Diersch, 2002), which is purported to have a conduit flow module with the capabilities needed in a karst aquifer modeling code. Initial testing suggested two different modeling approaches using MODFLOW may have the potential to replicate groundwater flow through a karst aquifer. Therefore, three modeling approaches were evaluated as part of this task: two using MODFLOW (one with modifications), and one using FEFLOW®.

Results from the 3-month evaluation suggested the standard version of MODFLOW was not adequate to perform the required simulations. Modifications to MODFLOW resulting in a dual-continuum representation, however, proved to be quite promising. One limitation of this modified version of MODFLOW (referred to as MODFLOW-DCM) was that simulations were limited to single-layer aquifers. The conduit package for FEFLOW® proved promising for confined aquifers, but inadequate for unconfined systems. These results were summarized at the interim project meeting held on April 5, 2004 at Marshall Center at the University of South

Florida. There were 24 participants at the meeting, including the core modeling team, representatives from the participating institutions, and representatives from the funding agencies.

Four separate modeling options were identified by the core modeling team. These options are listed in Table 1.1.

Table 1.1
Path forward options considered at the interim project meeting

Option 1	Finalize a single-layer version of MODFLOW-DCM
Option 2	Develop a multilayer version of MODFLOW-DCM
Option 3	Continue evaluation of FEFLOW®
Option 4	Initiate development of a new conduit flow code

The core modeling team at Southwest Research Institute® reported favorably on Options 1 or 2. The Technical Advisory Committee and the Steering Committee recommended that Option 1 be pursued in the first year. Both Committees recommended that Option 2 be considered in the later years of the project, if multilayer capability is required. The Steering Committee requested all code modifications be developed under Southwest Research Institute® Quality Assurance and Quality Control practices. The Steering Committee noted this effort should be completed by December 15, 2004, so a usable tool can be made available as a product of the first year of this karst modeling project. The product was specified to include documentation and validation of the code with descriptions of all analytical and semi-analytical solutions used in the validation process. Discussions during the Steering Committee meeting, clarified by subsequent discussions after the meeting, led to the recommendation that MODFLOW-2000 be the initial source code version for this task.

The second half of the first year was devoted to finalizing the DCM package for MODFLOW in two dimensions (2D), as well as testing, validating, and demonstrating the package. Three benchmark problems with independent numerical solutions were developed for use in the validation procedure. These validation exercises test the major capabilities of the DCM module and were successfully completed. The MODFLOW-DCM was used to model the Barton Springs Segment of the Edwards Aquifer near Austin, Texas as part of the demonstration process. The Barton Springs Segment has confined and unconfined zones and exhibits the classic characteristics of karst aquifers including sinkholes, sinking streams, groundwater troughs, and rapid spring response to recharge. Initial results with the DCM model applied to Barton Springs were encouraging. In particular, the DCM model was able to reproduce the rapid spring response and transient groundwater troughs.

STRUCTURE OF THE TECHNICAL REPORT

This report is a summary of the first year of the AwwaRF karst modeling project. Chapter 2 discusses the classification schemes for karst systems as related to groundwater flow modeling. Chapter 3 is a summary of modeling considerations for karst systems, including representations of conduits, and flow in conduits and considerations for unconfined karst aquifers. Chapter 4 describes three numerical experiments designed to test the performance of existing modeling approaches. Chapter 5 presents the basis for the DCM package for MODFLOW, which is designed to model aquifers within a dual-conductivity framework. Validation exercises and numerical experiments are also included in Chapter 5. The DCM model is applied to the Barton Springs Segment of the Edwards Aquifer in Chapter 6. Chapter 7 provides a summary, conclusions, and a discussion of future research needs.

CHAPTER 2
KARST PROPERTIES AFFECTING GROUNDWATER FLOW

The primary objective of this project is the development of general computational tools that can be applied to a wide range of karst aquifer conditions. The potential range of physical properties and flow behaviors is large, and developing computational tools applicable to all systems in this range is challenging. The first step in this process is to understand the range of flow behaviors expected in carbonate karst aquifers and how these behaviors relate to physical properties of the aquifers. A brief summary of relevant classification schemes for karst systems is presented, followed by a discussion of relevant flow regimes.

KARST CLASSIFICATION SYSTEMS

Karst aquifer systems have been characterized using diverse classification systems, depending on the ultimate use of the classification. Klimchouk and Ford (2000a) discuss the wide range of geologic environments from which a karst system can evolve. They note karst environments can be classified according to (i) structural prerequisites for groundwater flow and speleogenesis, (ii) flow regime, (iii) recharge mode and recharge/discharge configurations, (iv) groundwater chemistry, and (v) degree of inheritance (degree to which new features or conditions depend on previous conditions). In the context of hydrogeology, karst demonstrates a specific system of fluid circulation capable of self-development and self-organization because of its capability to evolve through the dissolution and transport of matrix minerals.

Klimchouk and Ford (2000b) developed a karst classification system based on lithologic and structural controls. These controls evolve with time and change drastically throughout sedimentation, eogenesis (during which sediments are laid down and subject to periodic drying and exposure), mesogenesis (during which sediments are buried and compacted), and telogenesis (during which rocks are uplifted and exposed to meteoric waters). Lithology may vary from carbonates (limestone and dolostone) to evaporites (gypsum, anhydrite, and salt). Porosity can vary widely in these rocks from as high as 80 percent in diagenetically immature carbonates to less than 2 percent in micrite and even lower in some evaporites and metamorphosed carbonates such as marble. Klimchouk and Ford (2000b) include bedding planes and contacts, joints and shear fractures, and faults in a broad class of structural guides for groundwater flow through karst rocks.

Palmer (1991) devised a frequently cited classification system based on morphology. Palmer's system identified four categories: spongework and ramiform, branchwork, anastomotic, and network. Palmer's analysis of thousands of caves indicated branchwork caves account for 65 percent, network caves account for 17 percent, anastomotic caves account for 10 percent, and spongework and ramiform caves account for 8 percent of caves based on total length passage. Palmer's analysis further indicated 57 percent of cave passages form along bedding planes, 42 percent form along fracture planes, and 1 percent are associated with intergranular porosity.

Anastomotic passages are phreatic features that closely follow parent bedding planes in rocks with high initial porosity. These passages tend to form in limestones that have not experienced significant tectonic fracturing and have a dip less than 5 percent. Anastomosing flow pathways initially form in tiny tubes with minimal flux under laminar flow conditions. The tubes enlarge with time, thereby reducing the hydraulic gradient. Palmer (1981) calculated that hydraulic gradient is inversely related by the fifth power to the tube radius. Tubes that enlarge tend to pirate flow from adjacent tubes developing depressions in the potentiomentric surface leading to increased tube discharge and increased pirating of flow.

Calic-Ljubojevic (2002) evaluated the effect of anastomosing networks of bedding-plane controlled conduits during the early stages of karst evolution. The corrosion processes associated with this type of passage tend to form in the beds above bedding planes rather than in the actual bedding planes. The incipient tubes in these planes start as circular openings, which may become elongated or elliptical in time. With progressive dissolution, the anastomotic conduit networks grow upward, often in multiple planes within a cyclic stratigraphic sequence. Collapse may occur after significant growth is experienced, however, there is no evidence that the passages cease to act as karst passages after collapse unless the formation undergoes mesogenesis.

Mylorie (2003) subdivided Palmer's classification into two groups: hypogenic (formed by deep processes), which includes spongework and ramiform caves, and epigenic (formed by shallow processes), which includes branchwork and anastomotic caves. Network caves are determined to form equally near the surface or at depth.

Kaufman and Braun (1999) developed a numerical code to simulate conduit evolution in a large-scale fractured karst aquifer. Their simulation results demonstrated that a fracture-controlled karst aquifer with fixed recharge developed a branchwork karst system. Systems with fixed hydraulic head boundary conditions and laminar flow developed a single passage karst system while fixed hydraulic head boundary conditions with turbulence developed a maze-like karst system.

Klimchouk (2003) associated cave morphology with confined versus unconfined speleogenetic settings. A confined setting is defined here to be a karstic stratigraphic bed sandwiched between nonkarstic units. These two cave categories demonstrate differences in passage network density (the ratio of cave length to the area of the cave field), cave porosity (fraction of the volume of a cave block occupied by mapped cavities), and average areal coverage (fraction of the area of the cave field occupied by passages in plan view). Of greater relevance to modeling karst systems is that conduit permeability in unconfined systems tends to be highly heterogeneous compared with homogeneous conduit permeability in confined systems.

Cave development is not necessarily restricted to one form of speleogenesis. Because cave development is dependent on discharge, CO_2 content, temperature, and thermal gradient, factors that favor one form of speleogenesis may vary with time, thereby instigating another form of speleogenesis (Palmer, 1991). The Guadalupe Mountains in New Mexico provide an example of a sequence of events that has changed cave development over 250 million years (Hill, 2000).

Shuster and White (1971) classified springs into two categories of feeder-system types, diffuse and conduit, based on dissolved carbonate species. The best indicator of a feeder-type system proved to be seasonally determined variation in groundwater hardness. Conduit springs were highly variable in hardness with a coefficient of variability of 10–20%. Conversely, diffuse flow springs exhibited little variation in hardness, typically having a coefficient of hardness <5%.

Karst aquifer systems may be characterized, or at least correlated, by age. Budd and Vacher (2004) characterized mean matrix, fracture, and channel permeability for 12 carbonate aquifers of ages Silurian to Quaternary. Matrix permeability was determined using core samples. Following Worthington et al. (2000), fracture permeability was determined using packer and slug tests, and channel permeability was determined using pump tests and regional-scale models. Mean values of the measurements compiled by Budd and Vacher (2004) are graphically summarized in Figure 2.1. As illustrated, there is an apparent correlation between age and permeability, with lower permeabilities observed in older rocks. These trends are slightly more evident in matrix and channel permeabilities than in fracture permeabilities, although the data used to establish the trends are limited, and the ages are approximated.

GROUNDWATER FLOW REGIMES AND RELATIONSHIP TO KARST PROPERTIES

As apparent from the previous discussion, karst systems exhibit a wide range of properties and may be classified in several different ways. For the purposes of groundwater modeling, the focus is on those properties that directly affect groundwater flow. Karst aquifer properties that have been identified as important for karst modeling studies include matrix permeabilities and storage properties, conduit storage, conduit size, conduit density, and conduit friction factors. Additional factors associated with recharge (i.e., epikarst and if the system is allogenic or autogenic) also are important.

Of particular importance for groundwater modeling are those properties that directly affect the spring discharge hydrographs. The dynamic response of a karst system to recharge is reflected in the spring discharge hydrograph. Replication of a spring discharge hydrograph with a model simulated spring hydrograph is a necessary (but not sufficient) requirement if the model is to adequately represent the dynamic hydraulics of a karst aquifer system. Inspection of spring discharge hydrographs from a wide variety of geological provinces indicates a wide range in general characteristics. Spring discharge hydrographs vary from extremely *flashy,* as indicated by a rapid increase in discharge quickly followed by a rapid return to a base flow level, to *subdued,* as indicated by a subtle, gradual increase in spring discharge followed by a gradual return to the base flow level. In particular, the degree of flashiness of spring discharge in response to recharge events is considered an indictor of the hydraulic dynamic character of a groundwater basin. A matrix with sufficiently large permeability has the potential to attenuate the flashiness of a karst system. Specifically, in karst systems with a highly permeable matrix, water is exchanged easily between the conduits and matrix, thereby attenuating hydraulic pulses in the conduits. This effect results in spring hydrographs that are more subdued than in systems with low matrix permeability. Karst systems, such as those found in the Florida platform, exhibit this latter behavior. An example is Silver Spring, located in the Floridan aquifer (Figure 2.2). Alternatively, karst systems with a matrix of low permeability do not significantly attenuate a hydraulic pulse traveling through the aquifer. This type of karst aquifer system will exhibit a flashy spring discharge hydrograph in response to a recharge (precipitation) event. Paleozoic karst systems, such as the Aruese Spring in the Jura Mountains, typically exhibit this behavior, as illustrated in a recorded hydrograph (Figure 2.2).

Although spring discharge flashiness can be correlated with geologic age, a more meaningful correlation is between spring discharge and diagenesis. For example, geologically young carbonate units that have undergone telogenesis (sediments that have experienced burial, compaction, and ultimately uplift) may exhibit flashy spring discharge common to karst carbonates of Paleozoic or Mesozoic age, even though they may be younger. Similarly, the spring discharge of older karst carbonates that have not undergone telogenesis may appear more subdued than otherwise expected. This perspective of karst diagenesis has been simplified for the purpose of classifying karst systems as numerical models.

This previous discussion is predicated on a simple recharge system. Compounding this simple approach are the effects of recharge. Recharge to a karst aquifer can be moderated or complicated by the character of the surface recharge. Intense precipitation events can be significantly subdued if the epikarst system acts as a buffer to infiltrating waters, causing recharge to be diffuse. Alternatively, if meteoric water is directly input into a karst aquifer, the discharge response at a spring will be intense. Once again, however, the recharge flux can be buffered in certain types of epikarst systems.

This analysis is also predicated on the assumption that the recharge basin is sufficiently simple to provide one discharge event at the spring. Hess and White (1988) analyzed the 340-km^2 Turnhole Spring subdrainage basin in south-central Kentucky that exhibits three distinct discharge features at Owl Cave. The three features represent (i) storm runoff from karst valleys, (ii) surface tributary drainage from an escarpment, and (iii) large-volume recharge contributed by a group of sinking streams. Discharge was characterized using measurements of specific conductance. Specific discharge measurements may exhibit a lag in arrival times relative to observed water discharge responses, consistent with the discharge of water in storage prior to the arrival of fresh precipitation. The most unambiguous response data were provided by intense, abrupt rain events of limited duration that occurred after relatively long periods (i.e., 2–3 weeks) of no precipitation.

In summary, classical karst classification schemes are not necessarily appropriate for categorizing karst systems as they affect groundwater flow. An alternative approach is to simply classify the karst systems according to the behavior of spring discharge hydrographs (i.e., whether spring discharge is flashy or subdued). Clearly, a continuum of possible discharge hydrographs is to be expected, depending on the matrix properties, conduit properties, and recharge characteristics.

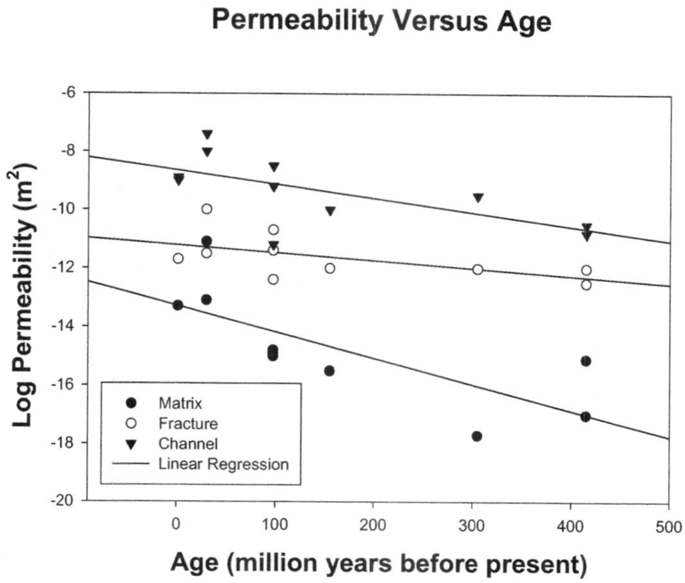

Figure 2.1 Permeability of carbonate rocks versus age (Data from Budd and Vacher, 2004)

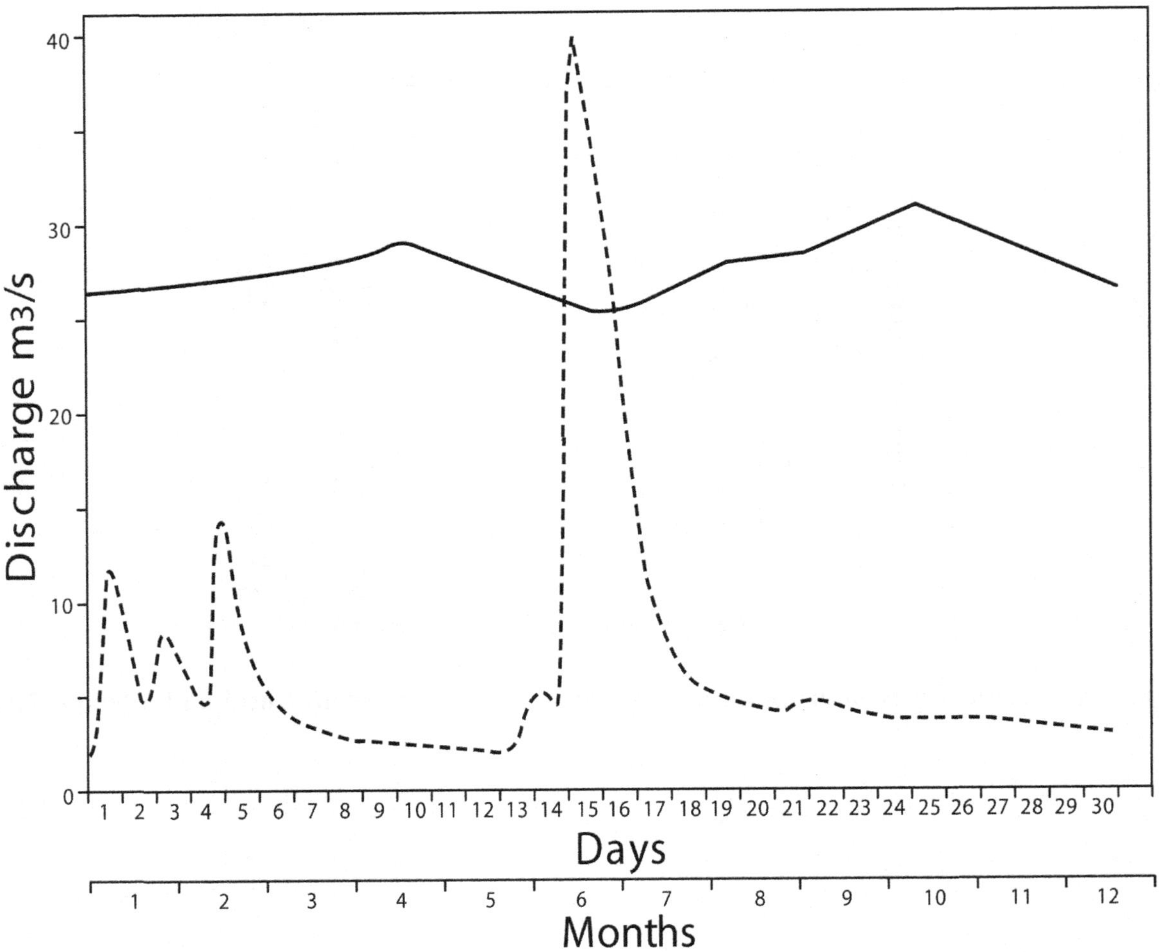

Figure 2.2 Discharge observed at springs located in different geologic environments. The solid line represents the Silver Spring, Florida, as an example of a subdued spring discharge system (taken from Faulkner, 1976). The dashed line represents the Areuse Spring in the Jura Mountains as an example of a flashy spring discharge system (taken from Kiraly, 2003a). The subdued spring response varies over 1 year and the flashy spring response varies over 1 month.

CHAPTER 3
KEY CONSIDERATIONS IN DISTRIBUTED PARAMETER MODELING OF KARST AQUIFERS

The selection of an appropriate modeling approach depends not only on the characteristics of the aquifer being modeled, but also on aquifer management needs. Predictions of solute transport may require, for example, quite different modeling approaches compared with an application addressing large-scale aquifer behavior for water supply considerations. In general, applications involving large spatial scales and global water budget considerations are more likely to be treated adequately with classical porous media concepts, whereas higher resolution models, or those addressing contaminant transport, are more likely to require more advanced modeling approaches.

This report addresses flow in *mature* karst aquifers with well-developed conduit networks. A *mature* karst aquifer, as suggested by Kiraly (2003b), is characterized by a high-permeability conduit network, immersed in a low permeability limestone volume, that is well connected to a local drainage area. The existence of conduit networks minimally enhances aquifer porosity, but greatly enhances aquifer permeability (Atkinson, 1977; Worthington, 2003). Worthington et al. (2000) gathered data on several carbonate aquifers where extensive caves have been found. Their analyses showed that conduits occupy only between 0.004 and 0.48% of the bedrock volume, however, more than 90 percent of the flow is through conduits. Thus, it is important to be able to model the flow in conduit networks and the mass exchange mechanism between conduits and the porous matrix. The velocity contrast between the least permeable and the most permeable parts in a channeled aquifer is often 6 to 10 orders of magnitude (White and White, 2003).

The strong heterogeneity and dual flow nature of mature karst aquifers introduces some special challenges for groundwater modeling. This chapter addresses three primary questions related to modeling karst aquifers: How do we represent conduits in groundwater models? How do we represent flow in open conduits? What are the unique considerations for unconfined karst aquifers?

REPRESENTATION OF CONDUITS IN GROUNDWATER MODELS

Flow in karst systems can occur in at least three modes. Matrix flow, or flow in the (primary) intergranular pore space, is laminar and generally accepted to be well described by Darcy's law. Laminar flow through secondary openings (joints, fissures, faults, and bedding plane partings) also occurs in many carbonate aquifers. The secondary and matrix flows can be lumped together and referred to as diffuse flow. Conduit flow occurs along solution features and may be laminar or turbulent, depending on the conditions.

Two possible conceptualizations of the conduit system are useful (Figure 3.1). The conduits can be thought of as pipes or considered as narrow zones of high transmissivity. In the former approach, each conduit is modeled as an equivalent pipe with a given cross section, transmissivity, and hydraulic radius. These properties may vary along the conduit. Mass exchange between the pipe and the diffuse system is proportional to the flow wetted area and is not controllable independently. In the latter approach, flow is conceptualized as occurring in a limited number of conduit zones. Each zone may contain multiple conduits. Rather than resolve each of these conduits, the entire zone is modeled as an equivalent highly transmissive region.

Karst aquifers have been modeled in the past using response function approaches, lumped parameter models, and distributed parameter models. The response function and lumped parameter models do not provide information on potentiometric surfaces and are generally limited to the predictions of spring flows. Only the distributed parameter models are discussed in this report. Distributed parameter models can be classified as embedded discrete-conduit models; single-continuum, smeared-conduit models; and dual-conductivity models. These three major types of distributed parameter karst models are illustrated in Figure 3.2.

Single-Continuum, Smeared-Conduit Models

Smeared-conduit models do not attempt to model detailed geometry of relatively small conduits. Instead, this approach attempts to capture the *effects* of a conduit on a much larger computational grid block by assigning effective properties to that grid block. In particular, grid blocks intersected by a conduit are assigned a large value for hydraulic conductivity, effectively smearing the conduit across the larger grid block. The approach can be implemented in a finite-element or finite-difference code, but is primarily applied to structured finite-difference codes. MODFLOW can be used unmodified to implement a single-continuum, smeared-conduit model, and several examples exist of this approach applied to the modeling of karst aquifers (e.g., Scanlon et al., 2003).

The smeared-feature approach has also been used for modeling flow in fractured rock, and the accuracy of the approach has been analyzed in some detail (Svensson, 2001), at least for steady-state conditions. Experience from fractured rock modeling suggests the smeared-feature approach is adequate for steady-state analysis.

However, questions remain about the accuracy of smeared-conduit models applied to transient simulations. In particular, accurately modeling transient flow requires the conduit cells be assigned relatively small values for storage properties. This assignment makes it possible to represent flashy behavior in the conduit system, but storage in the diffuse system is effectively ignored for those cells intersected by a conduit. It is not clear if interactions between the fast-responding conduits and the slower-responding diffuse system are adequately represented by relatively large computational cells that combine the small-volume, high-permeability conduits with the diffuse flow system.

Embedded Discrete-Feature Models

Compared with the finite-difference method, the finite-element method is better suited for representing complex geometrical shapes. Automatic mesh generation and mesh adaptation are offered by most modern finite-element software. In addition to offering better geometry representation, mesh adaptation allows creation of locally refined meshes at regions where the highest accuracy is desired, while keeping coarser discretization elsewhere in the model. This capability for constructing irregular finite-element meshes makes embedded discrete-conduit models possible.

Embedded discrete-conduit models are the most literal representations of conduits within a groundwater flow model. The basic approach is to explicitly represent each conduit within the finite-element mesh, with mesh refinement in the vicinity of each conduit. The conduits themselves may be represented as special one-dimensional (1D) line elements or as 2D "ribbon" elements. For example, Kiraly (1998) represented high conductivity karst channel network by 1D elements embedded in a three-dimensional (3D) mesh representing the diffuse flow system. The commercial finite-element software FEFLOW® (Diersch, 2002) has capabilities for implementing the embedded discrete-conduit model.

Dual-Conductivity Models

In the single-continuum, smeared-conduit and the embedded discrete-conduit models, a single heterogeneous flow system is represented. An alternative is to model two overlapping and interacting flow systems, the dual-conductivity model. In this approach, the diffuse background system is modeled as a continuum. Coupled to that continuum, and exchanging fluid with it, is a network of conduits. If the conduit network is dense, the conduit system can be modeled as a continuum, and the approach is identical to the dual-continuum model that has long been used in fractured-rock modeling. For mature karst aquifers, a relatively small number of conduits may dominate the flow, in which case the conduit system is a sparse network and not a continuum. Linear exchanges between the conduit system and the diffuse system are assumed, irrespective of whether the conduit system is modeled as a sparse network or a continuum. In this sense, the general approach is closely related to the classical dual-continuum model originally developed for fractured rock (Barenblat et al., 1960).

The dual-conductivity model has been applied to karst modeling with the conduit system represented as a continuum (Teutsch and Sauter, 1998). The CAVE module (Clemens et al., 1996), developed for use with MODFLOW-96, represents a partial implementation of the dual-conductivity model with the conduit system represented as a discrete network. The CAVE module does not include transient capability for the conduit system and is limited to confined aquifers.

REPRESENTATION OF FLOW IN CONDUITS

In mature karst aquifers, conduit flow is typically in the turbulent regime. For example, Halihan et al. (2000) estimate 95–99 percent of conduits in the Edwards Aquifer have Reynolds numbers greater than 2,000, which represents the approximate threshold for onset of turbulent behavior. Based on this analysis, all conduits with diameters greater than a few centimeters, which presumably dominate flow, would be in the turbulent regime.

For turbulent flow, the familiar linear relationship between Darcy velocity and hydraulic gradient is not valid and is replaced by a nonlinear flow law. By analogy with flow in engineering systems, the Darcy-Weisbach equation is typically assumed for flow in pipes. The Darcy-Weisbach equation relates the macroscopic head loss, Δh, in a straight section of pipe to the flow velocity

$$\Delta h = ff \frac{L}{D_H} \frac{\bar{v}^2}{2g} \tag{3.1}$$

where L is the length of the pipe, D_H is the mean hydraulic diameter, $\bar{v}$ is the mean velocity in the pipe, g is acceleration due to gravity, and ff is the friction factor. For straight pipes, ff depends on the relative roughness ε of the pipe and on the Reynolds number Re. Graphical representations of this dependence can be found in standard engineering handbooks (e.g., Murdoch, 1996). For a Reynolds numbers greater than about 4,000, the dependencies are also well represented by the implicit Colebrook equation (e.g. Murdoch, 1996).

$$\frac{1}{\sqrt{ff}} = -2\log\left[\frac{\varepsilon}{3.7} + \frac{2.51}{\mathrm{Re}\sqrt{ff}}\right] \tag{3.2}$$

Gale (1984) and Halihan et al. (1998) used similar Reynolds-number-dependent models for friction factors in natural conduits.

For rough pipes, the friction factor becomes independent of Reynolds number. A friction factor independent of Reynolds number appears to be appropriate for natural rough-walled conduits.

Springer (2004) pointed out that real conduit passages are rarely well approximated as straight pipes, but instead have bends, constrictions, expansions, and contractions. In engineering systems, the head loss caused by such arrangements of components is usually estimated by summing empirically determined values for each component (e.g., Murdoch, 1996):

$$\Delta h = ff\frac{L}{D_H}\frac{\bar{v}^2}{2g} + C_{bends}\frac{\bar{v}^2}{2g} + C_{ec}\frac{\bar{v}^2}{2g} \tag{3.3}$$

where C_{bends} is an empirically determined coefficient accounting for head loss in all bends, and C_{ec} is a similar coefficient for cross-sectional expansions/contractions. Springer (2004) used detailed conduit geometry, flow loss coefficients from engineering handbooks, and Equation (3.3) to calculate head losses in a segment of the Buckeye Creek Cave in West Virginia. The model was then used to estimate discharge for a flood with known head loss estimated from high water marks recorded as silt lines on cave walls. Independent estimates of the discharge were not available for verification.

In most applications, detailed conduit geometry is not available, and direct calculation of head losses from conduit geometry is impractical. Instead, a lumped parameter is needed that can be inferred or calibrated to match spring flows. To this end, each term in Equation (3.3) has identical dependence on velocity. Thus, the effects of the various bends and cross-sectional variations can be grouped into an effective friction factor

$$\Delta h = ff_e\frac{L}{D_H}\frac{\bar{v}^2}{2g} \tag{3.4}$$

An analogous form is more convenient for use in distributed groundwater models:

$$q = -\frac{k_c}{\sqrt{\|\nabla h\|}}\nabla h \tag{3.5}$$

where q is the Darcy velocity, and k_c is an effective conductivity for the conduit. Jeannin (2001) used this form in modeling flow in the Holloch cave in Muotatal, Switzerland, and calibrated values of k_c to match observed discharges. Jeannin also converted effective friction factors reported by several authors to an equivalent k_c. Results for eight different studies generally cluster conductivity estimates in the range 1–10 m/s.

Equation (3.5) is the preferred equation for modeling turbulent flow in conduits. However, groundwater modeling codes are typically based on the Darcy equation. It should be recognized that a Darcy model can always be calibrated to match a turbulent model in steady state. Specifically, an effective conductivity can be selected as $k_{eff} = k_c / \sqrt{\|\nabla h\|}$, which yields the same flow as the turbulent model in steady state. In transient conditions, however, the hydraulic gradient will necessarily deviate from the value used in calibration, and the two flow models will diverge.

For applications involving transient simulations, it is important to understand the degree of error introduced by using a Darcy model to approximate turbulent flow. A simple numerical experiment was constructed using FEFLOW® to quantify this error. The flow configuration involves a single conduit embedded in a confined aquifer. The conduit is connected to a recharge point on the one end and a spring on the other end. See Experiment 1 in Chapter 4 for details of the model parameters and flow configuration. The recharge rate is 0.1 m^3/s. In the first run, it is assumed the conduit flow is Darcian; in the second run, it is assumed the conduit flow is turbulent, and the Darcy-Weisbach law[1] is used. In the first step, k_{eff} was calibrated such that the Darcian flow and turbulent flow yield the same steady-state spring discharge. In the second step, the system behaviors were simulated in response to the recharge event. In Figure 3.3, the transient spring discharges obtained by the Darcy and turbulent flow models are compared. For clarity, the base-flow is removed in both curves. It can be seen that a large deviation exists between results of the two flow laws, with the peak discharge of the Darcy model being three times greater than that of the turbulent flow model. Figures 3.4(a) and 3.4(b) show the steady-state and transient head distributions along the conduit. At steady state, the Darcy and turbulent flow models predicted similar head distributions along the conduit. At the transient state, however, the head at the conduit origin as predicted by the turbulent flow is 50% higher than that predicted by the Darcy model.

In conclusion, although appropriately calibrated Darcy flow and turbulent flow can yield similar results at steady state, for transient conditions, the two models will likely diverge because of the dependency of effective conductivity on the head gradient in the turbulent flow law. Thus, a turbulent flow model is necessary for modeling transient flow in karst aquifers with well-developed conduits.

CONSIDERATIONS FOR MODELING FLOW IN UNCONFINED KARST AQUIFERS

There are some specific modeling considerations associated with unconfined karst aquifers. If the water level drops below the top of a conduit, the conduit is then only partially filled with water, and transmissivity of the conduit also is reduced. Similarly, if the water level

---

[1] FEFLOW® does not offer the Darcy-Weisbach equation directly, but does have an option for Manning's equation. Manning's equation has dependence on the hydraulic gradient (or slope for open channel flow) identical to Equation (3.5), but is simply parameterized differently from the Darcy-Weisbach equation. Thus, the Darcy-Weisbach equation can be simulated using Manning's equation with appropriate selection of the parameters. In particular, $k_c = \frac{1}{n} A R^{2/3}$ where A is the conduit cross-sectional area, R is the hydraulic radius, and n is Manning's empirical coefficient.

drops below the base of the conduit, the conduit no longer has an effect on the flow in the aquifer. Thus, to model an unconfined karst aquifer, it is necessary to use information on the top and bottom of the conduit or conduit zone in the simulation, even if the groundwater model is a single-layer model. The MODFLOW CAVE module and FEFLOW® with the line-element option both model conduits as one-dimensional elements with no vertical extent. Thus, these modeling options are appropriate for confined systems but are, in general, not useful for modeling unconfined systems. Using the two-dimensional (ribbon) element option in FEFLOW® to represent conduits does allow the user to specify the vertical extent of conduits or conduit zones. However, in Version 5.0 of FEFLOW®, the transmissivity of the discrete features is calculated as if the conduit is filled with water. That is, adjustment is not made for saturated thickness of the conduit. In addition, Version 5.0 of FEFLOW® does not include storage in the discrete elements in unconfined mode. Thus, this option is also not useful for modeling unconfined aquifers with partially filled conduits.

Drying/rewetting of conduits is another computational consideration for unconfined karst aquifers. Because the vertical extent of conduits or conduit zones may be limited, relatively small changes in hydraulic head can cause the conduit to move from filled to completely dry. In MODFLOW, the iteration scheme used to resolve nonlinearities in the unconfined flow equations exhibits a well-known instability if the water level in a computational cell approaches the bottom elevation for that cell. This instability is a general disadvantage for MODFLOW. Preliminary tests indicate that this drying/rewetting instability is amplified by the presence of conduits and is a serious barrier to applications. Proposed modifications to the MODFLOW standard handling of drying cells, which are designed to alleviate this instability, are discussed in Chapter 5.

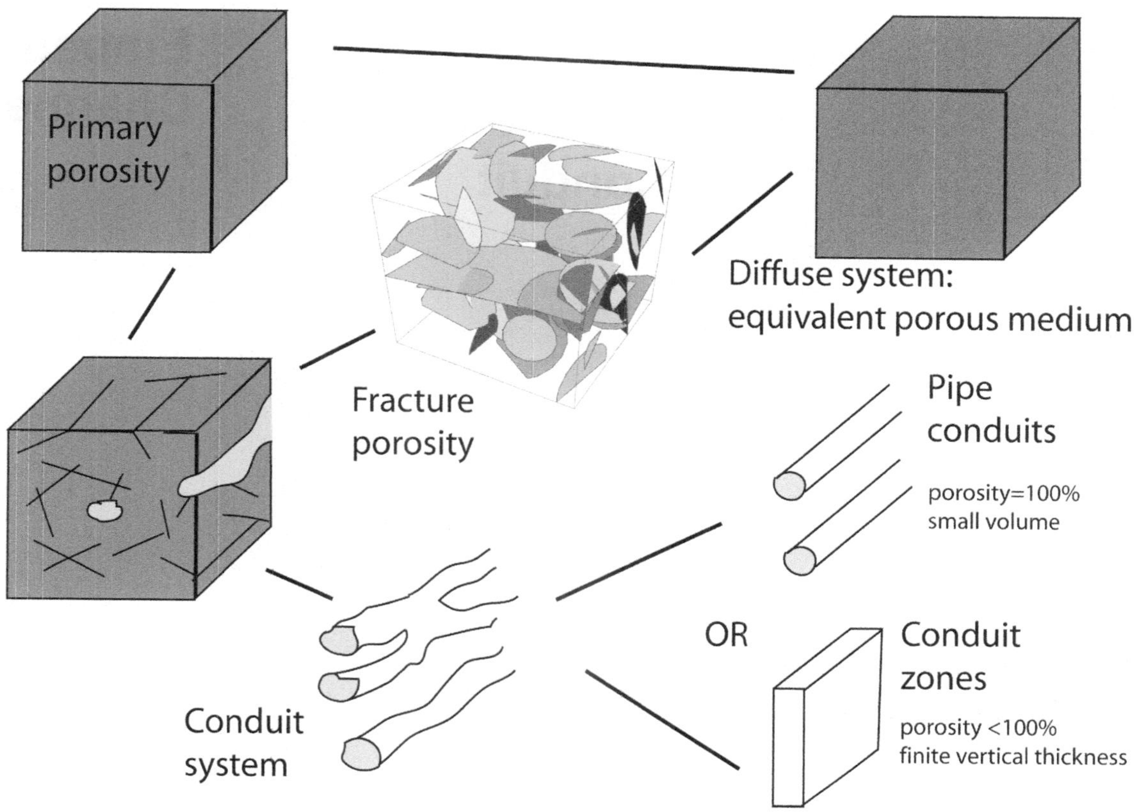

Figure 3.1 Flow conceptualizations for karst aquifers

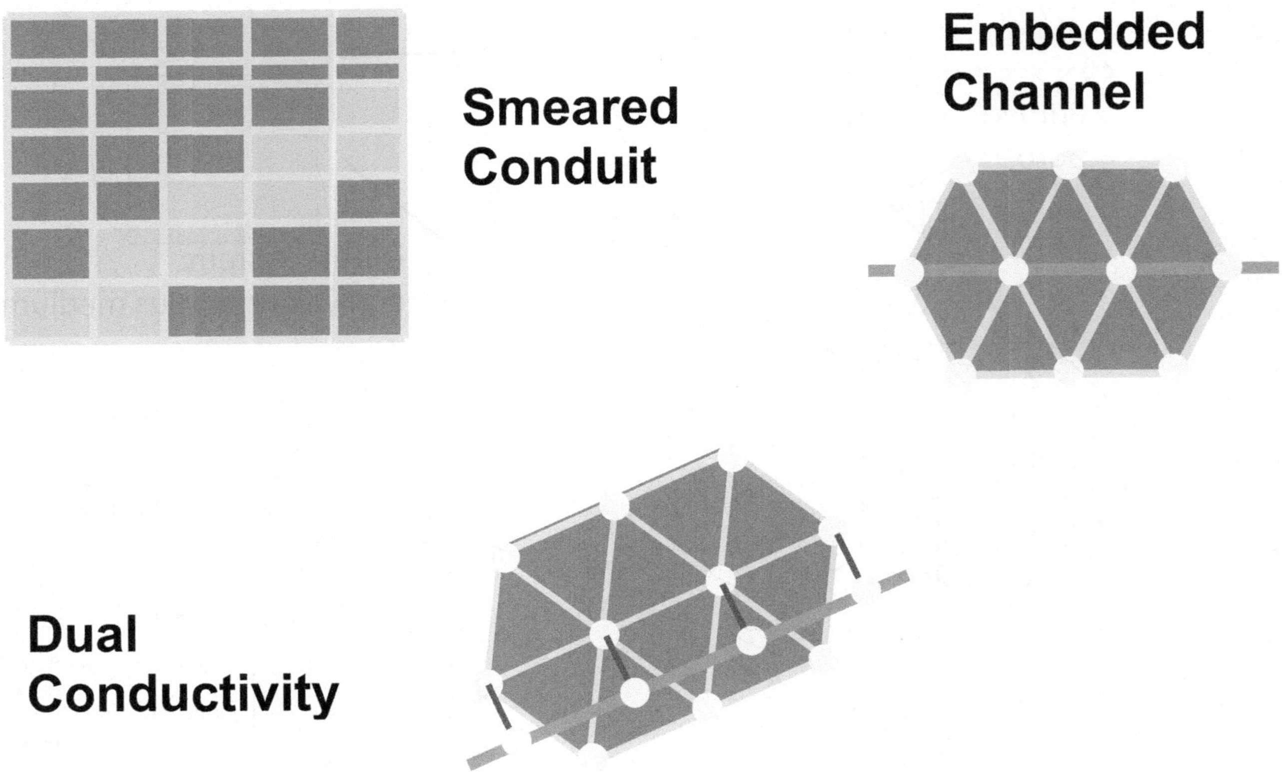

Figure 3.2 Three modeling approaches for representing effects of karst conduits in groundwater models

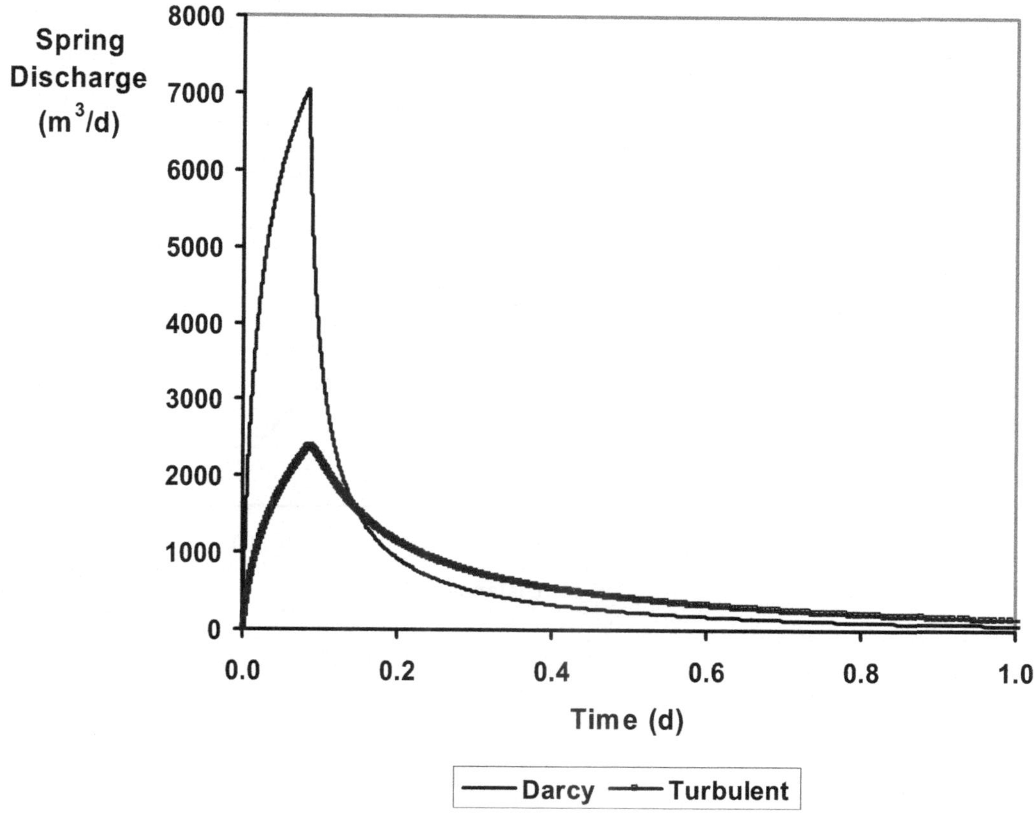

Figure 3.3 Spring discharge versus time predicted by Darcy and turbulent flow models. The two models are calibrated to provide similar steady-state discharges but produce very different transient behavior.

(a)

(b)

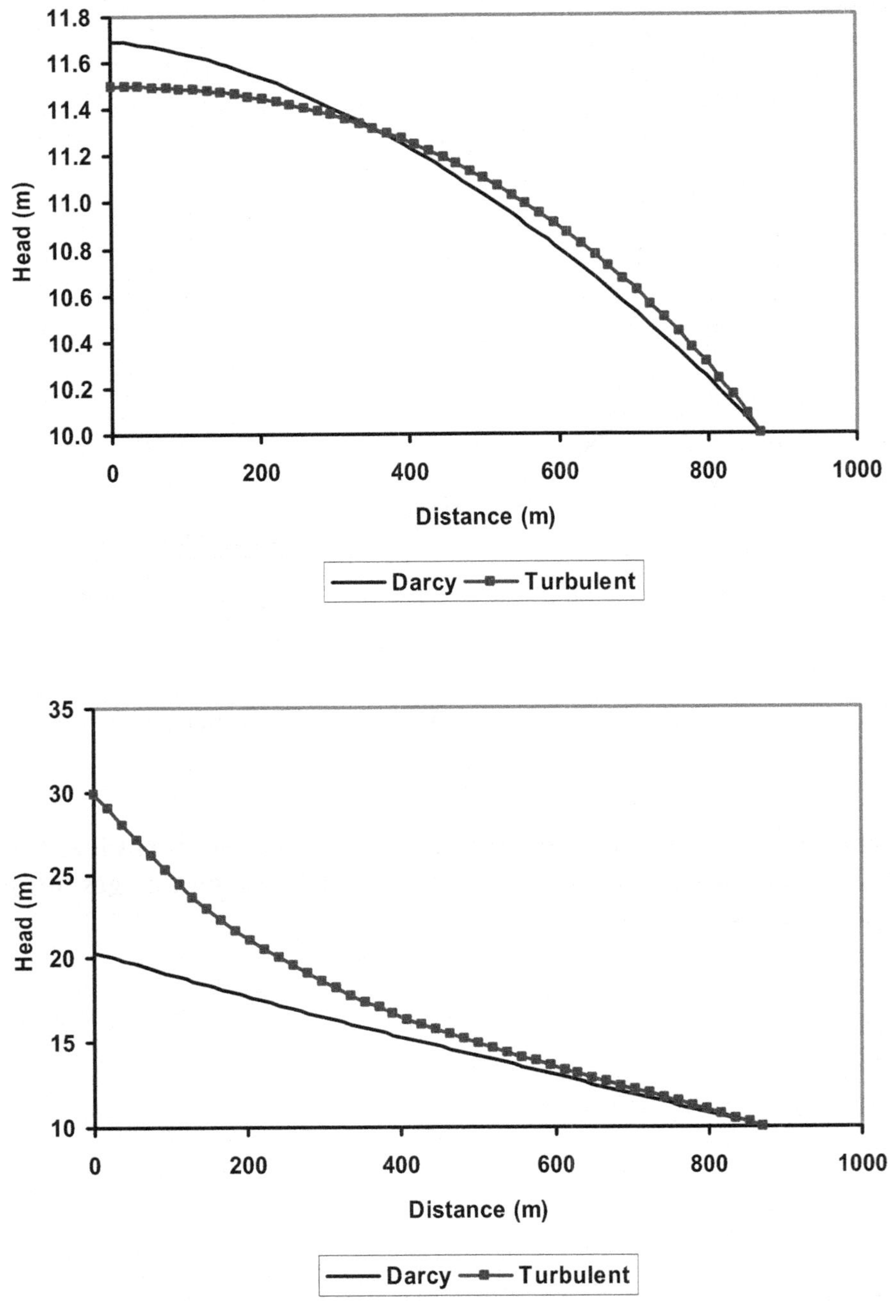

Figure 3.4 Hydraulic head variation along the conduit predicted by the Darcy and turbulent flow models (a) steady-state conditions and (b) at the end of the recharge event

CHAPTER 4
PERFORMANCE OF SMEARED-CONDUIT MODELS UNDER CONFINED CONDITIONS

A primary goal of the project is to evaluate the performance of karst modeling options. A series of numerical experiments was used to test the capability of smeared-conduit representations to approximate discrete conduit networks. The FEFLOW® and MODFLOW codes were used in the comparisons.

APPROACHES

About FEFLOW® and MODFLOW

FEFLOW® (Finite Element Subsurface Flow and Transport Simulation System) is a commercial, 3D, finite-element code (Diersch, 2002). This code can be used to simulate flow and mass or heat transport in variably saturated aquifers. The latest version, FEFLOW® 5.1, offers an additional capability for modeling discrete fractures or channels that are embedded in a porous medium.

MODFLOW (MODular groundwater FLOW) is a public-domain, 3D, finite-difference code (Harbaugh et al., 2000) for simulating groundwater flow in confined and unconfined aquifers. MODFLOW has been used widely by various government agencies and consulting companies for groundwater modeling projects.

Table 4.1 compares FEFLOW® and MODFLOW and lists the pros and cons. Some items revealed in the table will be addressed in the next sections.

Smeared-Conduit Approach

In a finite-difference-based regional groundwater model, it is generally impossible to delineate the actual shape of a conduit without incurring higher computational cost. Moreover, the precise positions and detailed shapes of conduits are generally not known. In practice, much coarser contiguous blocks are often used to approximate the geometry of the conduit. This is the *smeared-conduit approach.* Besides the difficulty in representing the actual geometry of a conduit, it is difficult to incorporate anisotropy into the smeared-conduit approach without special treatment. For a network consisting of a single straight conduit, one can always align the grid such that the conduit coincides with one of the major axes. For more complex conduit networks, however, there is no such option. United States Geological Survey (Anderman et al., 2002) recently added an LVDA (model-Layer Variable-Direction horizontal Anisotropy) module to MODFLOW-2000. LVDA allows the principal directions of horizontal anisotropy to be different from the grid alignment. LVDA also allows the principal directions to vary on a cell-by-cell basis within model layers. LVDA is limited to modeling variable direction anisotropy within a layer.

Table 4.1
FEFLOW® and MODFLOW comparisons

	FEFLOW®	MODFLOW
Model Dimension	2D/3D	2D/3D
Numerical Method	Finite Element	Finite Difference
Automatic Grid Generation and Adaptation	Y	N
Flow	Y	Y
Unconfined to Confined Aquifer Transition	In 2D, storage parameter depends on aquifer thickness, which varies with saturation. In 3D, a moving mesh technique is used to model free surface movement	Conversion from dry to wet is based on a heuristic method-a wetting threshold is used to form a cell. A cell is converted to wet if the head in its adjacent cell is higher than its wetting threshold, which can be problematic at times
Spring Modeling	A spring can be incorporated through a combinations of constant head boundary condition and flux boundary constraints	A spring can be modeled through the DRAIN module
Time-stepping Control	Either constant or automatic can be used	Constant
Fracture/Conduit Modeling	Through finite elements or through 1D/2D discrete-elements	Through fine cells
Fracture Flow Governing Law	Darcian flow (laminar), Hagen-Poiseuille flow (laminar), and Manning-Strickler flow (turbulent)	Darcian flow only
Mass and Heat Transport	Y	N
Nonlinearity	Newton's method with predictor/corrector timestepping or Picard iteration with constant timestepping	Picard iteration with constant timestepping
Preprocessing	Built in: offers connectivity to data generated by Geographic Information Systems	Through commercial preprocessors
Postprocessing	Built in: offers nice 3D graphics. Export capabilities in the current Budget Analyzer and Flux Analyzer are insufficient	Through commercial postprocessors
License	Proprietary	Public domain
Add-Ons	Users can add customized modules through programming the Interface Manager Module	The source code is free and, thus, is modifiable at the user's discretion

Discrete-Element Approach

Compared with the finite-difference method, the finite-element method is better suited for representing complex geometrical shapes. Automatic mesh generation and mesh adaptation are offered by most modern finite-element software. In addition to better geometry representation, mesh adaptation allows one to create locally refined meshes at regions where the highest accuracy is desired, while keeping coarser discretization elsewhere in the model.

For a complex model (e.g., a conduit network embedded in a 3D porous matrix), mesh adaptation can still introduce a large computational burden because of the number of regions that need to be refined. The discrete-element approach offers a way to alleviate such a problem. Numerically, this approach is nothing but a mesh having mixed element types, and there is only one set of discrete equations to be solved. With the discrete-element approach, however, the details of a conduit can be modeled, provided enough information is available. That is, the hydraulic parameters assigned to conduits are the intrinsic parameters. In addition, modeling anisotropy is easily accomplished in the finite-element method without worrying about grid orientation.

To summarize, the discrete-element approach attempts to model the detailed process; the smeared-conduit approach makes some compromise between model resolution and model precision, and the conduits and porous matrix are lumped together. Neither the discrete-element approach nor the smeared-conduit approach explicitly models the coupling between the conduit flow and the porous matrix. Given this fact and assuming that conduit data is not a constraint, it is interesting to ask how well the smeared-conduit approach performs. To make such a comparison, a means is needed to relate the parameters in the two models.

Assuming the FEFLOW® discrete-element model represents 'reality', the corresponding model parameters must be calculated for a smeared-conduit model, which is set up in MODFLOW. Although it is possible to set up a smeared-conduit model in FEFLOW®, MODFLOW was used because both the approaches and the performance of the numerical codes were evaluated.

Parameter Assignment for Smeared Conduits

Figure 4.1 shows a 3D prism that represents a computational cell taken from a smeared-conduit model. The conduit volume is only a small portion of the whole volume. Given the intrinsic specific storage and hydraulic conductivity for the conduit, a method is needed for assigning effective properties of an equivalent smeared-conduit computational cell.

The governing equation for flow in a confined aquifer is

$$S_s \frac{\partial h}{\partial t} = \nabla \cdot (\mathbf{K} \cdot \nabla h) + Q \tag{4.1}$$

where h [L] is hydraulic head, S_s [L^{-1}] the specific storage, $\boldsymbol{K}$ [L/T] the hydraulic conductivity tensor, and Q [T] the volumetric flux per unit volume representing a sink/source term.

Flow along the x direction, based on mass balance, presents the following relationship:

$$K_c A_c + K_m A_m = K^* A^* \tag{4.2}$$

where the subscripts c and m denote conduit and matrix, respectively; the superscript * stands for effective properties, and A represents the area. The total area A^* is the sum of A_c and A_m. Rearranging Equation (4.2) yields the effective hydraulic conductivity for the smeared-conduit cell:

$$K^* = \frac{K_c A_c}{A^*} + (1 - \frac{A_c}{A^*})K_m \approx \frac{K_c A_c}{A^*} = \frac{K_c A_c}{\Delta y \Delta z} \tag{4.3}$$

In the previous equation, it is assumed $A_c << A^*$ and $K_m << K_c$. Note that the effective properties of the cell dependent on the cell size. That is, the effective property of the smeared-conduit representation is not independent of the model resolution.

To obtain specific storage properties, set

$$S_{s,c} A_c \Delta x = S_s^* \Delta x \Delta y \Delta z \tag{4.4}$$

where $S_{s,c}$ is the specific storage of the conduit, so the volume of water liberated per unit head drop will be the same as in a pure conduit system. Thus, the effective specific storage for use in the smeared-conduit approach is

$$S_s^* \approx \frac{S_{s,c} A_c}{\Delta y \Delta z} \tag{4.5}$$

If flow is along a different axis (e.g., along the y direction), Δx and Δy need to be switched in Equations (4.3) and (4.5).

It can be seen from Equations (4.3) and (4.5) that the effective properties may be orders of magnitude smaller than the intrinsic conduit properties. It is emphasized that effective properties calculated from Equations (4.3) and (4.5) are not exact quantities; they are only used as the initial estimates for the smeared-conduit cells. In the actual numerical experiments to follow, additional calibrations will be performed to obtain the best match between the MODFLOW and FEFLOW® results.

NUMERICAL EXPERIMENTS

Three groups of numerical experiments were performed to test the performance of the smeared-conduit approach. Figure 4.2 shows the 2D-confined aquifer used in all tests. The domain is 1,000 × 1,000 m. The aquifer thickness is 10 m. The aquifer is bounded by constant heads on both the upper and lower boundaries. The head at the upstream boundary is 25 m, and the head at the downstream boundary is 15 m. The lateral boundaries of the domain are assumed to be no-flow boundaries. The conductivity of the background porous matrix is 1×10^{-4} m/s, and the specific storage is 1×10^{-4} m^{-1}.

Experiment 1

In this experiment, there is a single conduit in the aquifer. One end of the conduit is a spring, and the other end is connected to a recharge feature that receives external recharge in the simulation. The location of the spring is (100 m, 250 m), and the location of the recharge point is (875 m, 650 m). The simulation starts from a steady-state condition. A transient simulation begins with a 2-hour period of recharge at the specified recharge point. The recharge rate is 0.5 m^3/s.

Nodes used in the FEFLOW® model number 1,743, and the element type is triangular. To add the conduit to the model, a line is drawn in the direction of the conduit. FEFLOW® treats the line as an internal boundary and automatically generates refined mesh around the line. (Note that depending on the geometry, it is possible to control the total number of elements used and the degree of refinement in FEFLOW®.) Finally, discrete line elements are added on top of the line. Thus, a conduit is created with refined mesh around it. The value of hydraulic conductivity assigned to the conduit is 10 m/s, and specific storage is 0.005 m^{-1}. The cross-sectional area of the conduit is 1 m^2. The final finite-element mesh is shown in Figure 4.3(a).

The MODFLOW model has 1,600 cells. Each cell block has dimensions of 25 m × 25 m. To delineate the smeared-conduit, blocks were selected in the direction of the conduit, creating a zigzag shape. Figure 4.3(b) shows the MODFLOW grid. Cells that constitute the smeared conduit are highlighted with dark dots. The model is set up with the MODFLOW-2000 Layer-Property-Flow (LPF) package.

The calculated value of the smeared-conduit hydraulic conductivity is 0.04 m/s, and specific storage is 2×10^{-5} m^{-1} using Equations (4.3) and (4.5) for initial estimates. In the calibration process, the attempt was to match the MODFLOW steady-state spring discharge to that of FEFLOW® by adjusting the smeared-conduit hydraulic conductivity. In the second step, the attempt was to match the transient MODFLOW spring discharge to that of FEFLOW® by adjusting specific storage. The resulting calibrated smeared-conduit conductivity is 0.04 m/s, and specific storage is 2×10^{-7} m^{-1}. The smeared-conduit was assumed to be isotropic in this case.

A comparison of the hydraulic head distributions calculated by the two models along the horizontal line y = 362.5 m is shown in Figure 4.4. Figure 4.4(a) shows the steady-state head profile, and 4.4(b) shows the transient-state head profile taken at the end of the recharge event. From the figures, it can be seen that the smeared-conduit approximation matches the explicit conduit very well in steady state. The maximum deviation occurs near the left boundary and is about 0.5 m. In transient conditions, however, the smeared-conduit approach overestimates the head in the upstream portion of the conduit. The maximum deviation in head is almost 3 m. This deviation is an indication the smeared-conduit is less efficient in conveying water downstream, compared with the actual conduit.

The head distribution along the conduit is shown in Figure 4.5. For the smeared-conduit, the head values are taken from those cells that fall on the line connecting the spring and the recharge point. The origin of the horizontal axis corresponds to the recharge point (i.e., looking from upstream to downstream along the conduits). As shown in Figure 4.4, the smeared-conduit representation overestimates the hydraulic head in the upstream regions of the model.

The spring discharge hydrograph has been used as an indicator of the closeness of match in the calibration process. In practice, it is possible to gage a spring to obtain its discharge hydrograph data. Halihan et al. (1998) continuously measured velocity and stream stage near the discharge point of Devil's Icebox, a small karst drainage basin located near Columbia, Missouri. They calculated the spring discharge rate based on these data.

Spring discharge hydrographs given by the two models are compared in Figure 4.6. MODFLOW predicts smaller spring discharge during the recharge event. Maximum deviation occurs at the peak of the hydrograph and is approximately 15%. A smeared-conduit involves a larger contact area with the porous matrix than the actual conduit does. In addition, the effect of anisotropy is not captured in this analysis. These two factors may result in a greater mass exchange between the smeared-conduit and the porous matrix and, thus, a smaller outflow from the spring. The FEFLOW® spring hydrograph is calculated using its budget analyzer, which gives the outflow flux at the spring node. The MODFLOW hydrograph is calculated by multiplying the drain conductance and the head difference at the drain cell. Different handling of spring flow by FEFLOW® and MODFLOW may have contributed to the difference in the spring hydrographs. In MODFLOW, the spring discharge is averaged over a larger area than in FEFLOW®, which may reduce the spring discharge calculated by MODFLOW. Finally, the difference in numerical schemes (the FEFLOW® triangular element involves linear basis functions, which is first order; MODFLOW employs a center finite-difference scheme, which is second order) is difficult to quantify in the current case and may have also contributed to differences in discharge results.

Closely related to Experiment 1, two supplementary experiments were conducted. In the first supplementary experiment, the effect of anisotropy is assessed. Using the LPF cell-by-cell horizontal anisotropy feature, the smeared-conduit cells are made anisotropic, and the anisotropy ratio is based on the orientation of the conduit. For Experiment 1, this ratio is (650–250m) divided by (875–100m), which is 0.516. The resulting spring hydrograph is shown in Figure 4.7. Incorporating anisotropy, alone, dramatically improved the comparison to this experiment.

In the second supplementary experiment, the effect of mesh refinement level was assessed. FEFLOW® offers no refinement, medium refinement, and high refinement around any kind of internal boundary. This experiment showed that, for the single-conduit configuration adopted in Experiment 1, the effect of mesh resolution on calculated spring discharge hydrograph is minimal.

Experiment 1 shows the smeared-conduit approach satisfactorily captures the dynamics of the simple karst conduit system. The concern is whether or not the smeared-conduit approach will break down in more complex cases as the volume of conduits becomes larger. This is the task for the following two experiments.

Experiment 2

In Experiment 2, a simple conduit network is used to test the robustness of the smeared-conduit approach. The boundary conditions and hydraulic parameters are the same as those used in Experiment 1. The simple conduit network consists of two branches, which drain to the same spring. The recharge is split between the two branches (i.e., 0.25 m^3/s for each recharge point). Figure 4.8 shows the grids used in Experiment 2.

Head distributions along a horizontal line drawn at $y = 762.5$ m are compared in Figure 4.9. MODFLOW consistently overestimates the head values. Figure 4.10 shows the predicted spring discharge hydrographs. Again, the smeared-conduit approach satisfactorily captures the dynamics of this simple conduit network system. Because the aquifer is confined, the effect of the travel time difference in the two branches on the discharge hydrograph is hardly observable.

Experiment 3

A more complex conduit network is used in this experiment. Figure 4.11 shows the grids used in this experiment. Nine branches comprise the network; the total recharge rate of 0.05 m^3/s is partitioned equally among the branches. As seen in Figure 4.11, a larger portion of the upstream of the system is occupied by the conduits. The intrinsic conduit conductivity in FEFLOW® is set to 1 m/s, and the calibrated smeared-conduit conductivity in MODFLOW is found to be 0.004 m/s. The specific storage values are the same as those used in Experiment 1.

Spring discharge hydrographs are shown in Figure 4.12. Compared with FEFLOW®, MODFLOW predicts an earlier peak time and a lower recession curve. These differences can be attributed to the fact that, in the smeared network, a larger volume is taken by the low-storativity smeared continuum, which causes the system to have more "flashy" behavior during an infiltration event. Figure 4.13 shows the head distributions along a cross section drawn at $y = 387.5$ m. Although there is a good match between head distributions near the spring, MODFLOW over predicts the head distribution significantly in the upstream. This difference cannot be calibrated out as long as the spring discharge is used as the calibration criterion. In other words, it is impossible to calibrate simultaneously the head distribution and the spring discharge with the current set of parameters. Finally, color maps of head distributions are shown (Figure 4.14) for the middle of the recharge event. The MODFLOW result shows larger head values around the recharge points.

Experiment 3 demonstrates that the performance of the smeared-conduit approach deteriorates as the total volume of conduits becomes larger. A simultaneous match to the hydraulic heads and spring discharges in transient simulation cannot be achieved by calibrating the single-continuum, smeared-conduit properties.

SUMMARY

In this chapter, performance of FEFLOW® and MODFLOW are compared. The finite-element method is better suited to represent complex geometries. In practice, many consulting companies and government agencies prefer MODFLOW because of their investment and effort in setting up existing MODFLOW models. At regional scales, an actual conduit usually cannot be explicitly represented by a finite-difference model, such as a MODFLOW model. Instead, a modeler has to resort to coarse grids in which a block cell includes a mixed volume of conduit and porous matrix. This is the smeared-conduit approach. A scheme for estimating smeared-conduit properties based on intrinsic conduit properties was proposed and performance of the smeared-conduit approach was tested with a series of three numerical experiments.

These experiments show the smeared-conduit model represents the behaviors of simple conduit networks relatively well. As the karst network structure becomes more complex and the conduit density becomes larger, performance of the smeared-conduit approach becomes poor.

Refining the grid resolution would likely improve performance of the smeared-conduit approach; however, that would also result in a globally refined grid with a large computational burden. Although there are methods available for making a locally refined mesh in MODFLOW, such as the telescopic mesh refinement method (Leake and Claar, 1999), these methods introduce additional complexities. With such local refinement, the relative simplicity of the MODFLOW approach is lost.

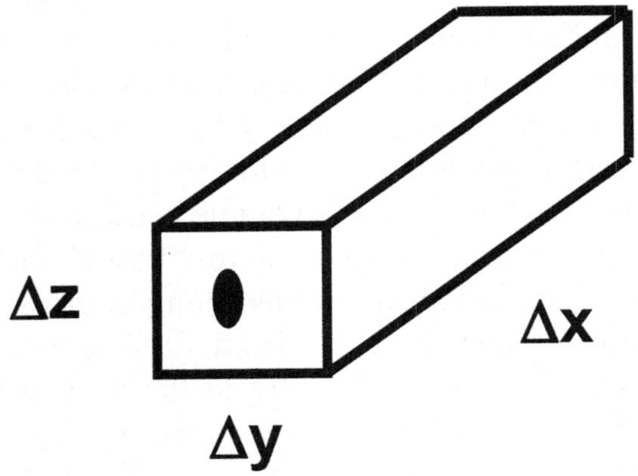

Figure 4.1 Plot of the 3D prism used in calculating smeared-conduit parameters

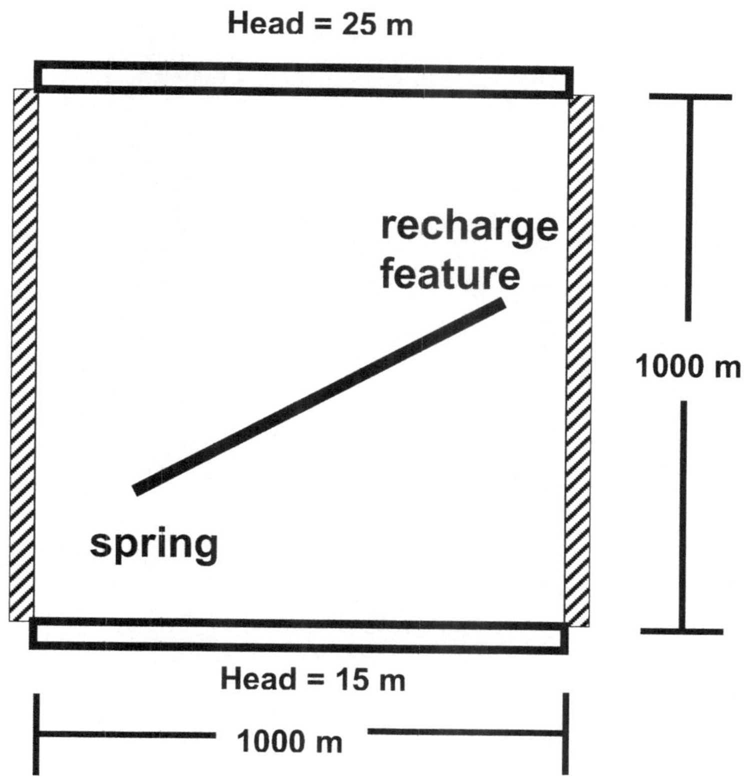

Figure 4.2 Configuration of the 2D confined aquifer used in numerical experiments

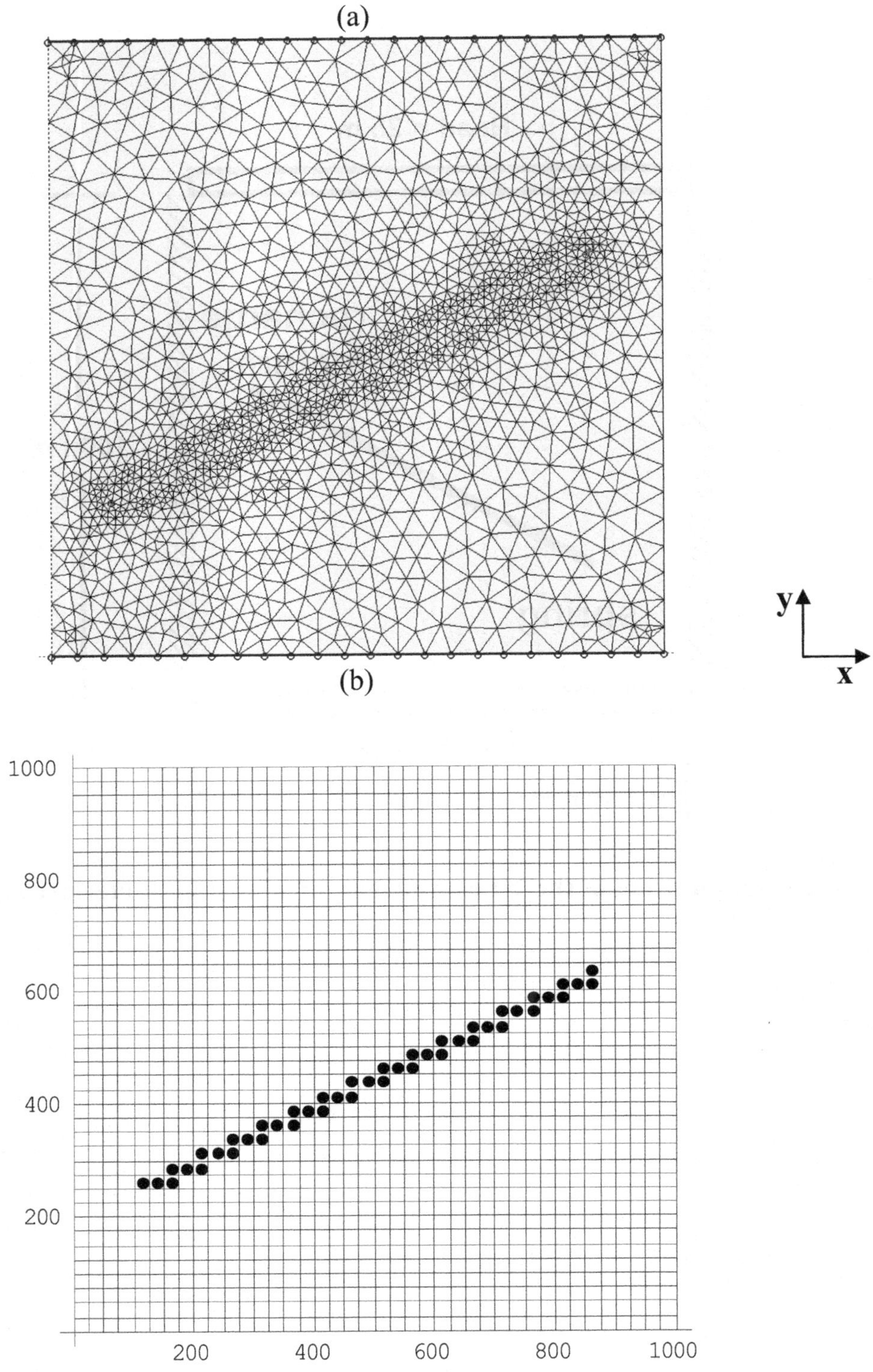

Figure 4.3 Grids used in numerical Experiment 1. One end of the conduit is a spring (100 m, 250 m), and the other end is a recharge point (875 m, 650 m). (a) The FEFLOW® mesh, and (b) the MODFLOW grid.

(a)

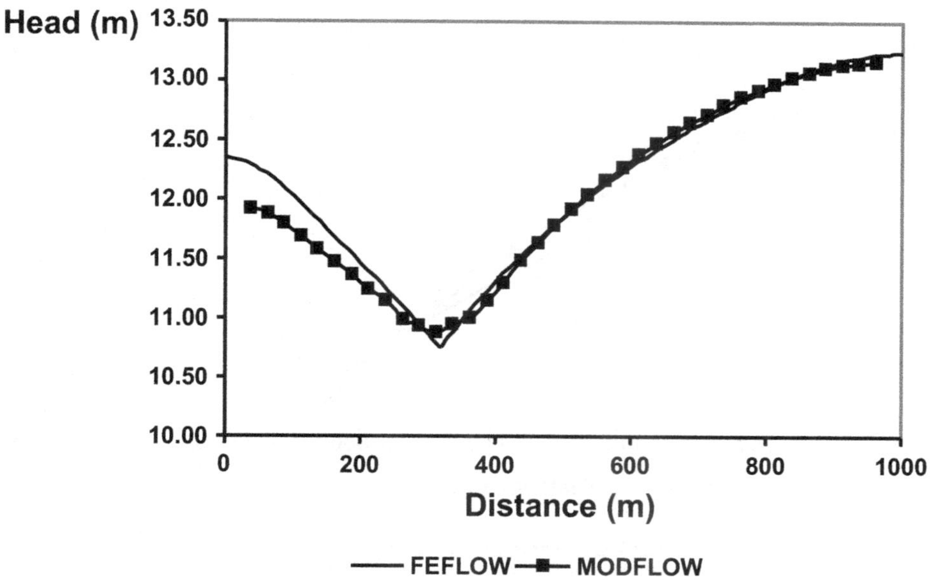

(b)

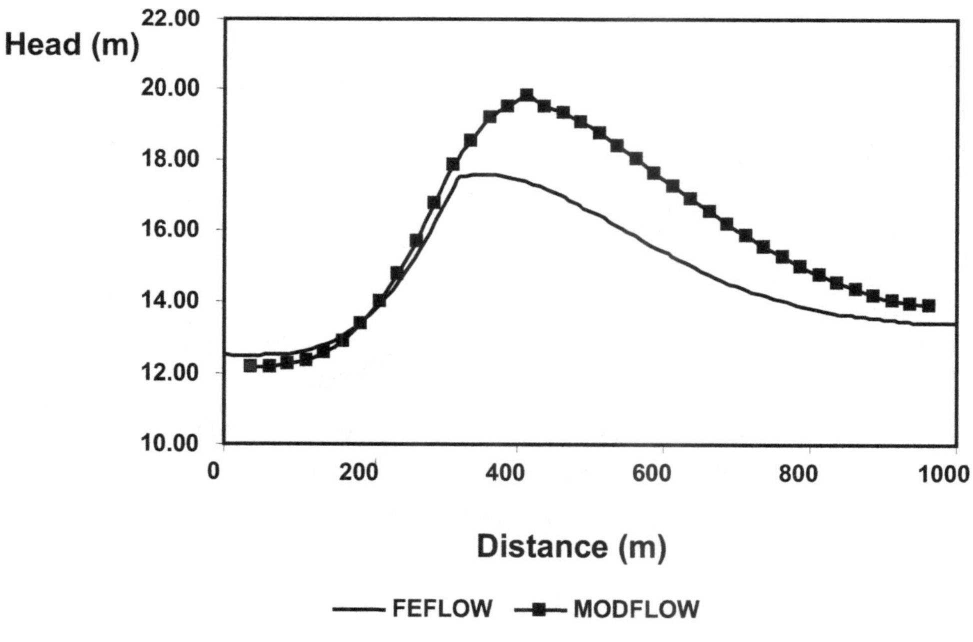

Figure 4.4 Comparison of the head distributions along a horizontal line drawn at y = 362.5 m in numerical Experiment 1. The solid line is the FEFLOW® result, and the line with squares is the MODFLOW result. (a) The steady-state head distribution profile and (b) the head distribution at the end of the recharge event.

(a)

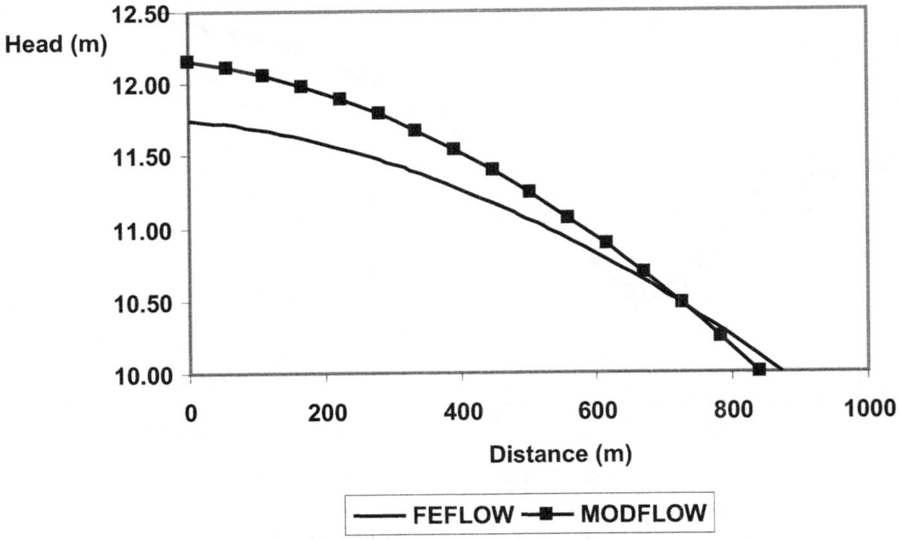

(b)

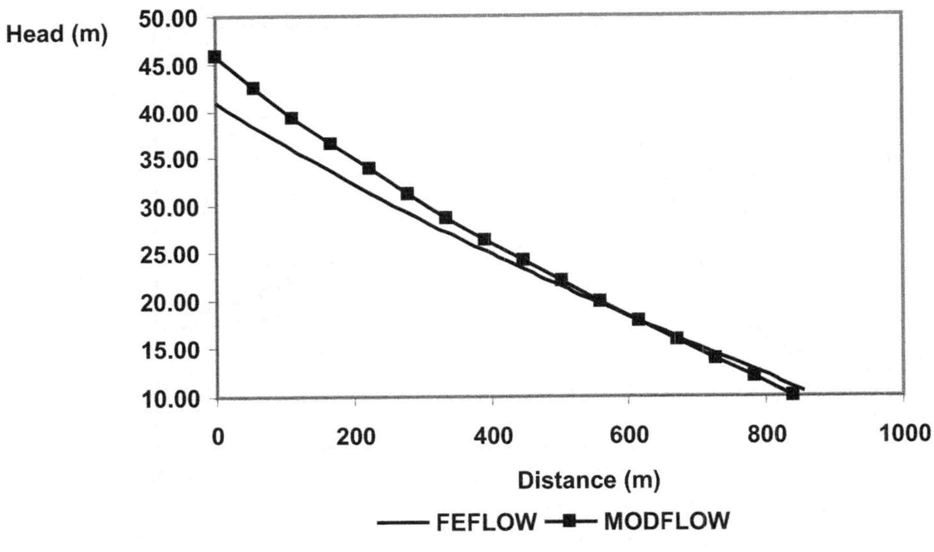

Figure 4.5 Comparison of the head distributions along the conduit in numerical Experiment 1. FEFLOW® result is shown in solid line, and MODFLOW result is shown in squares. (a) The head distributions at steady state, (b) the head distributions at the end of the recharge event.

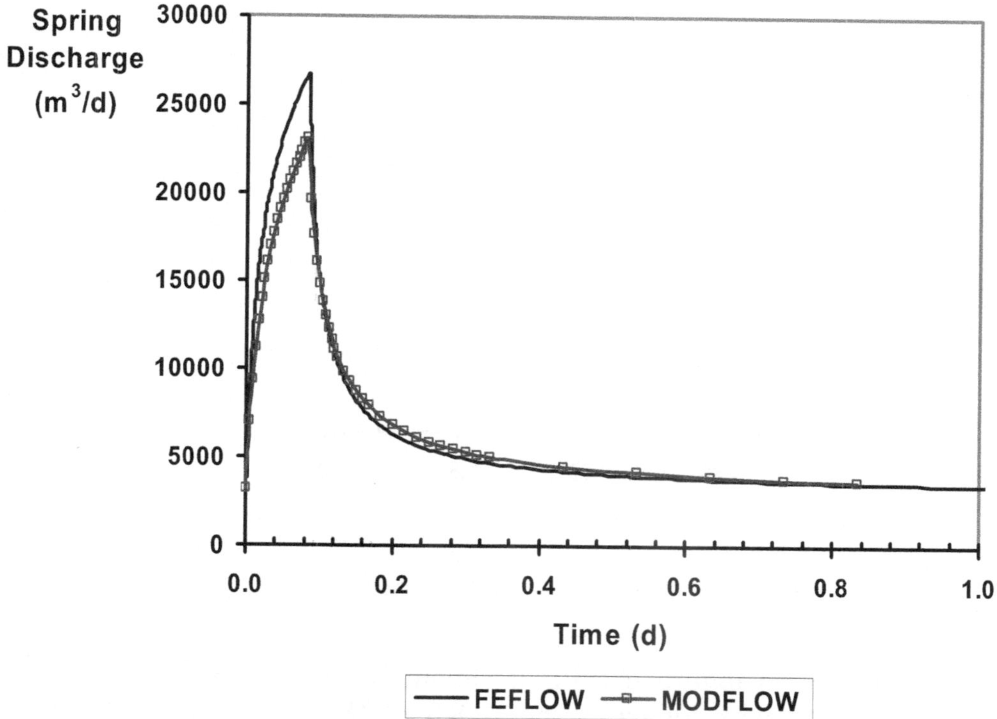

Figure 4.6 Comparison of the spring discharge hydrographs in numerical Experiment 1. The solid line is the FEFLOW® hydrograph, and the line with squares the MODFLOW.

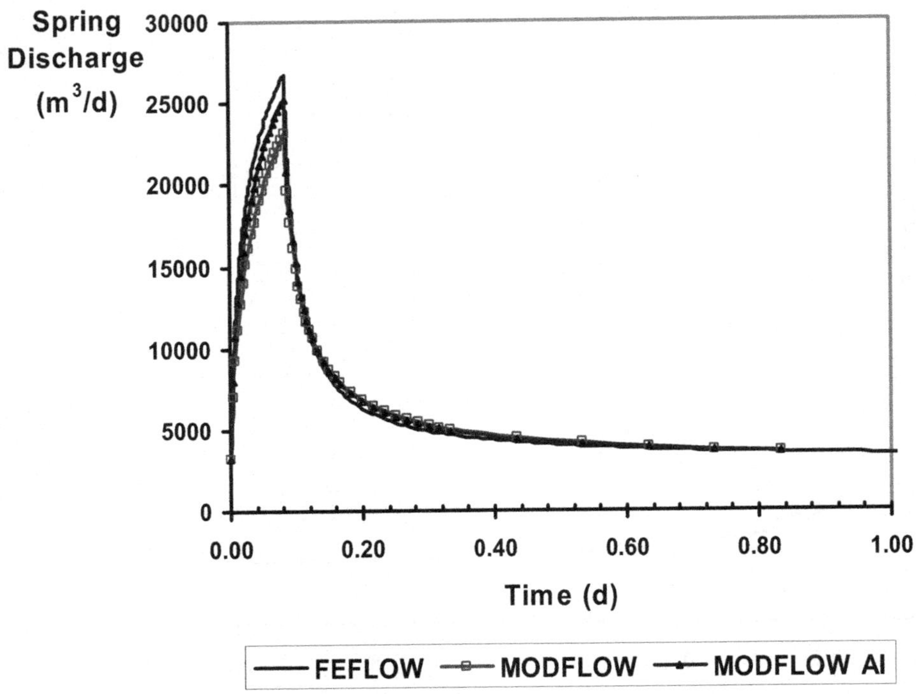

Figure 4.7 The effect of anisotropy in numerical Experiment 1. The solid line is the FEFLOW® hydrograph, the line with squares is the MODFLOW result based on isotropic conduit conductivity, and the line with triangles is the MODFLOW result for anisotropic conduit conductivity.

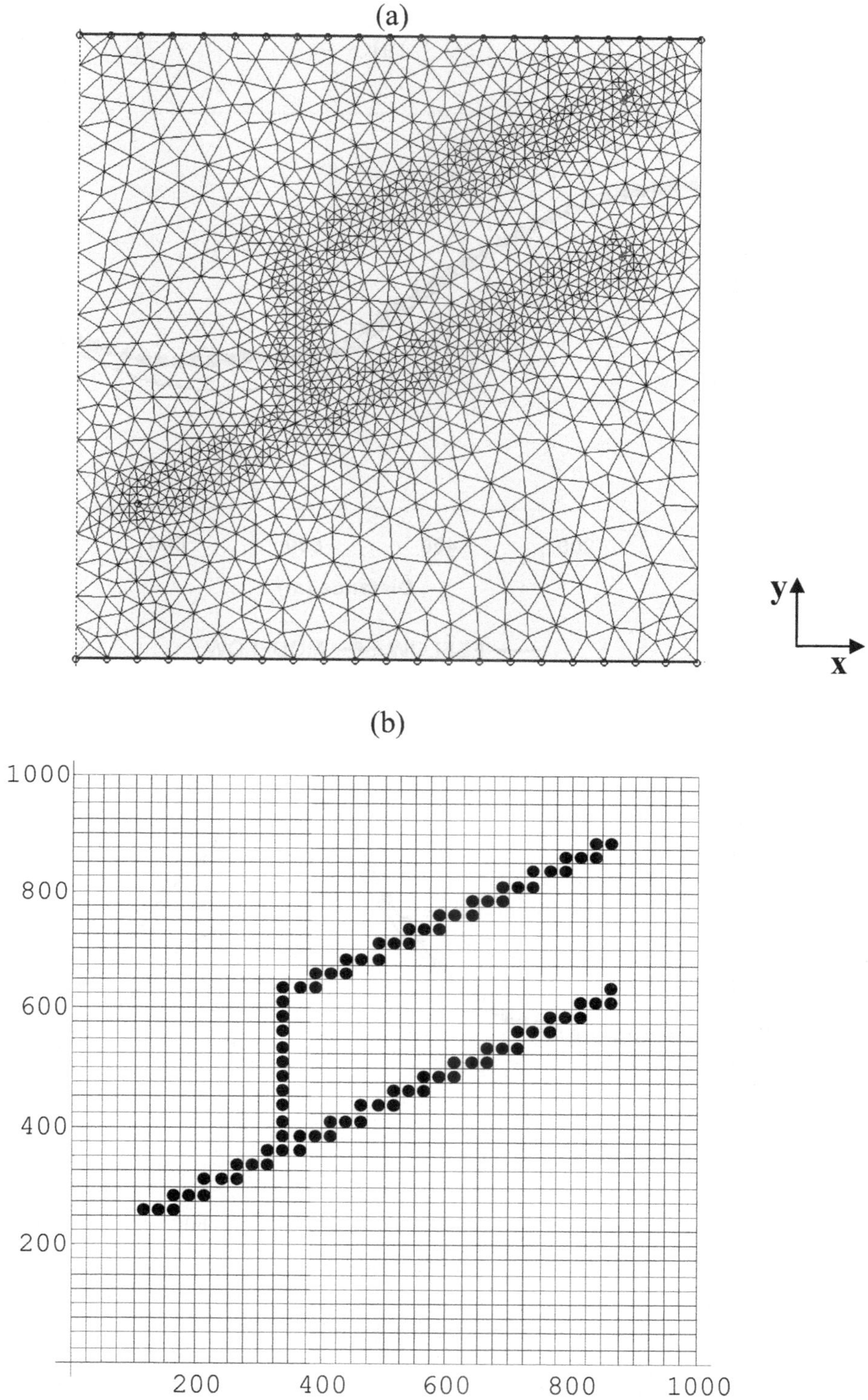

Figure 4.8 Grids used in numerical Experiment 2. (a) The FEFLOW® mesh, and (b) the MODFLOW grid

(a)

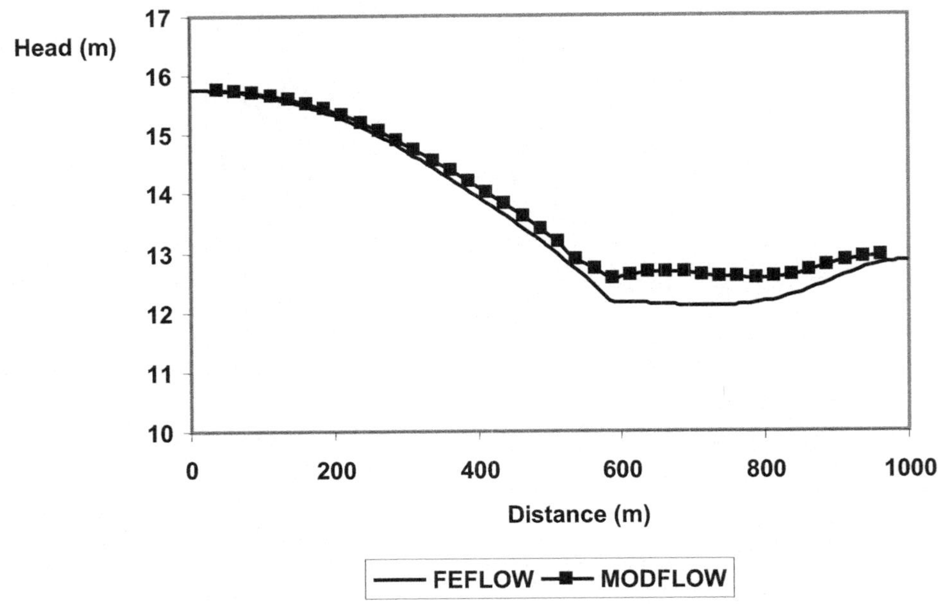

(b)

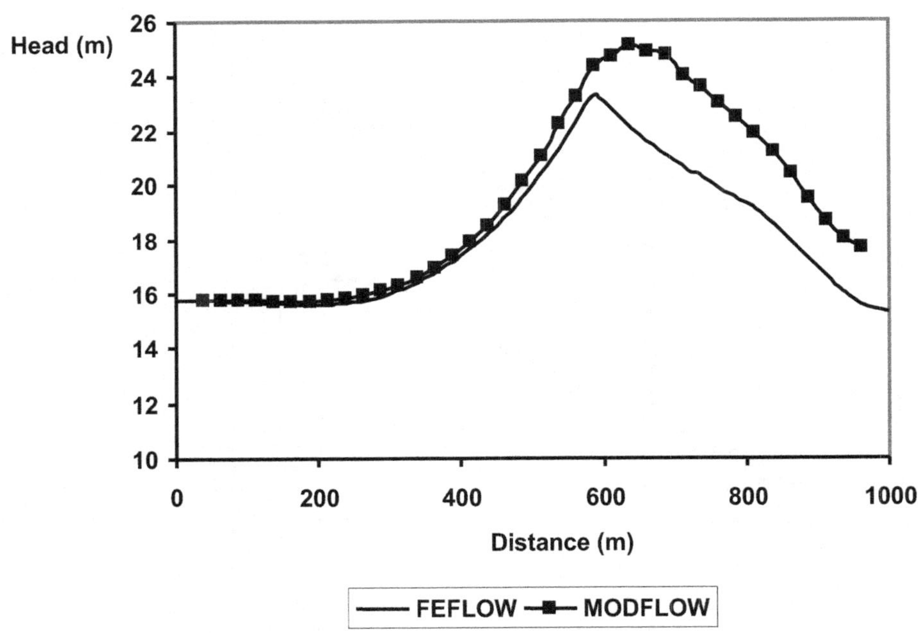

Figure 4.9 Head distributions along a section drawn at y = 762.5 m in numerical Experiment 2. (a) Comparison at steady state, and (b) comparison at the end of the recharge event.

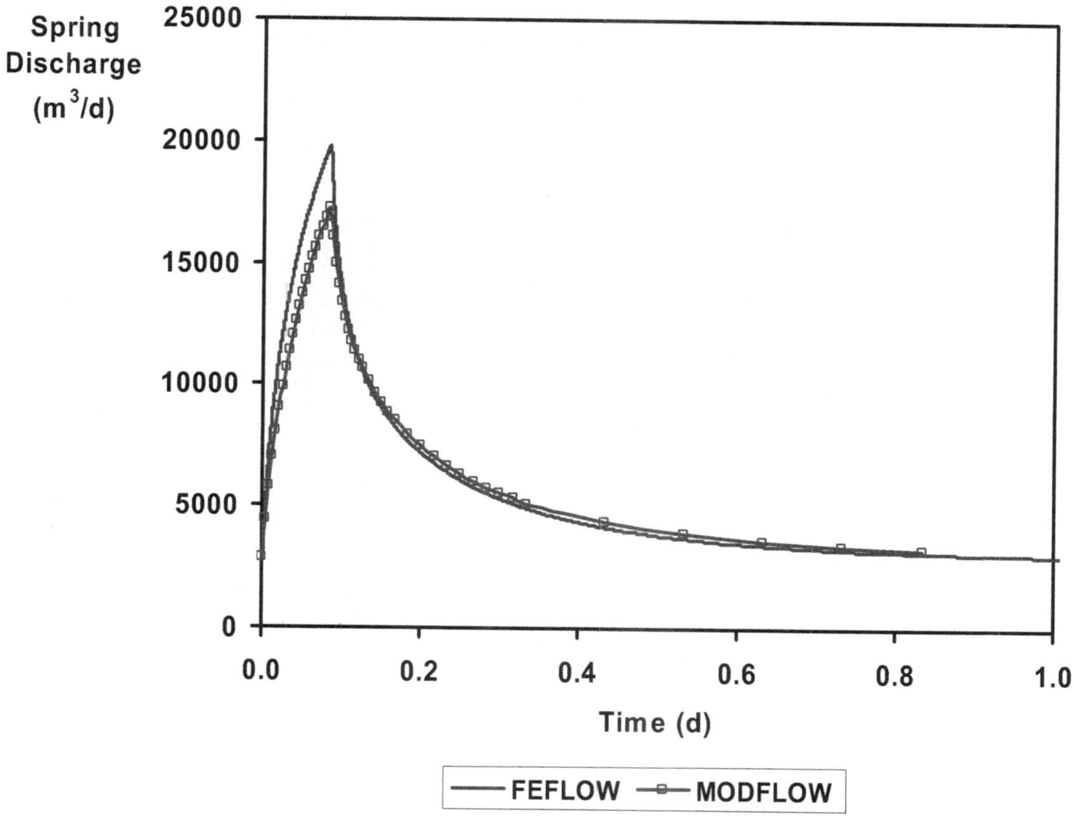

Figure 4.10 Comparison of the spring discharge hydrographs for numerical Experiment 2. The solid line represents the hydrograph generated by FEFLOW® and the line with squares is the hydrograph generated by MODFLOW.

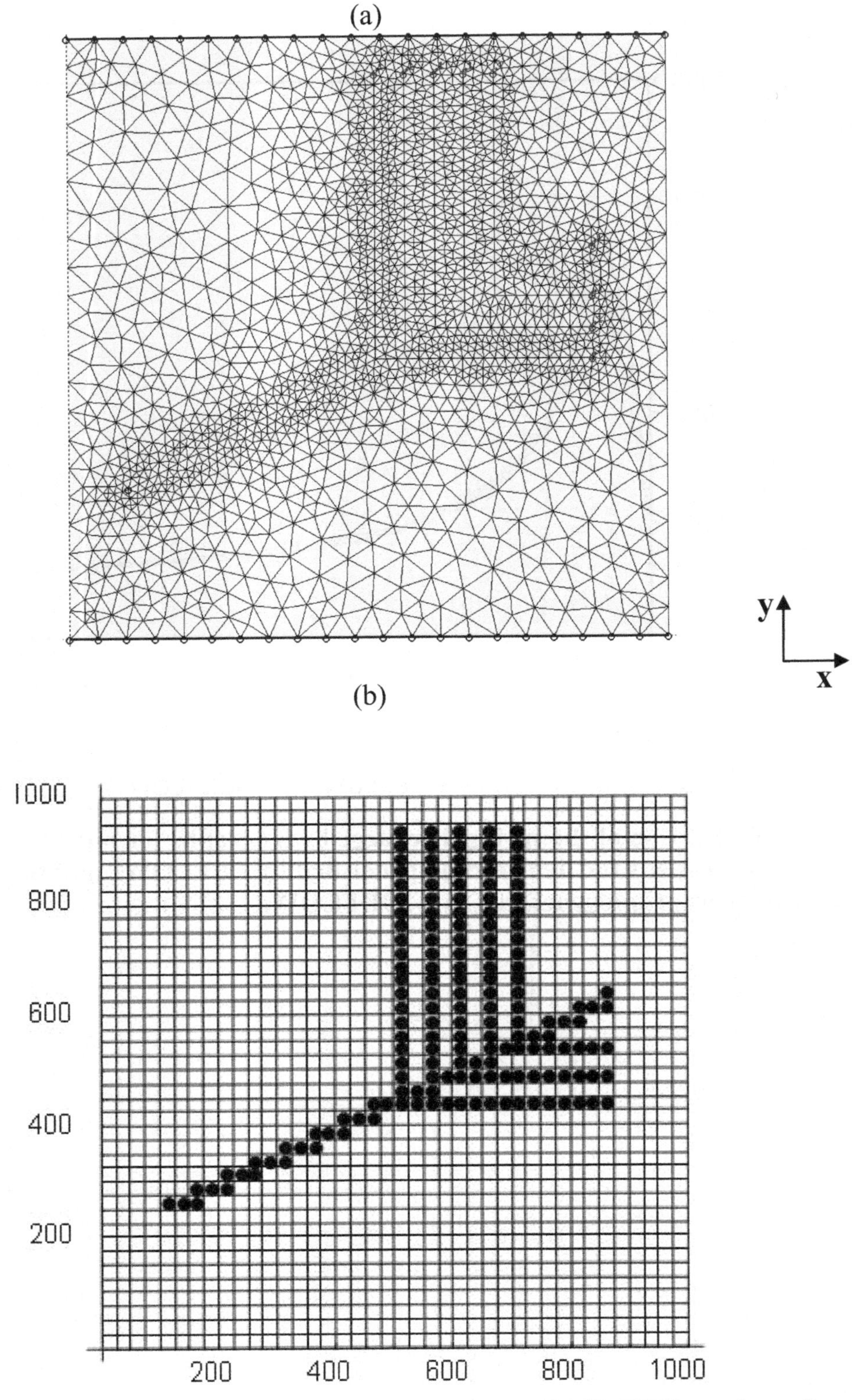

Figure 4.11 Grids used in numerical Experiment 3. (a) The FEFLOW® mesh, and (b) the MODFLOW grid.

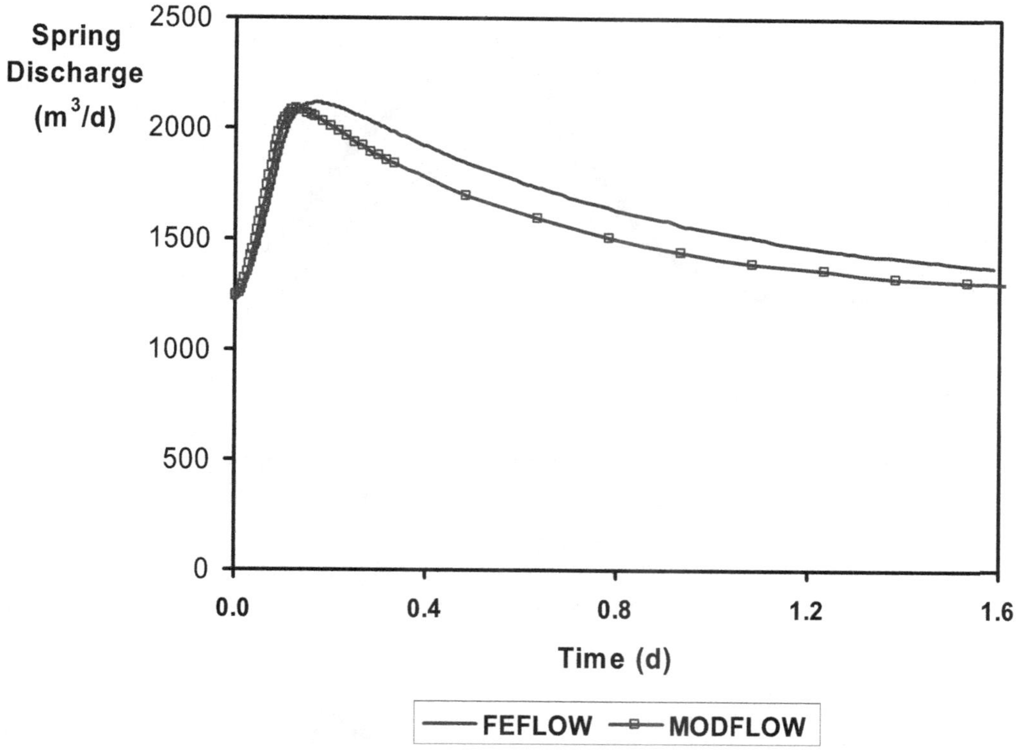

Figure 4.12 Comparison of the spring discharge hydrographs for numerical Experiment 3. The solid line represents the FEFLOW® hydrograph and the line with squares is the hydrograph generated by MODFLOW.

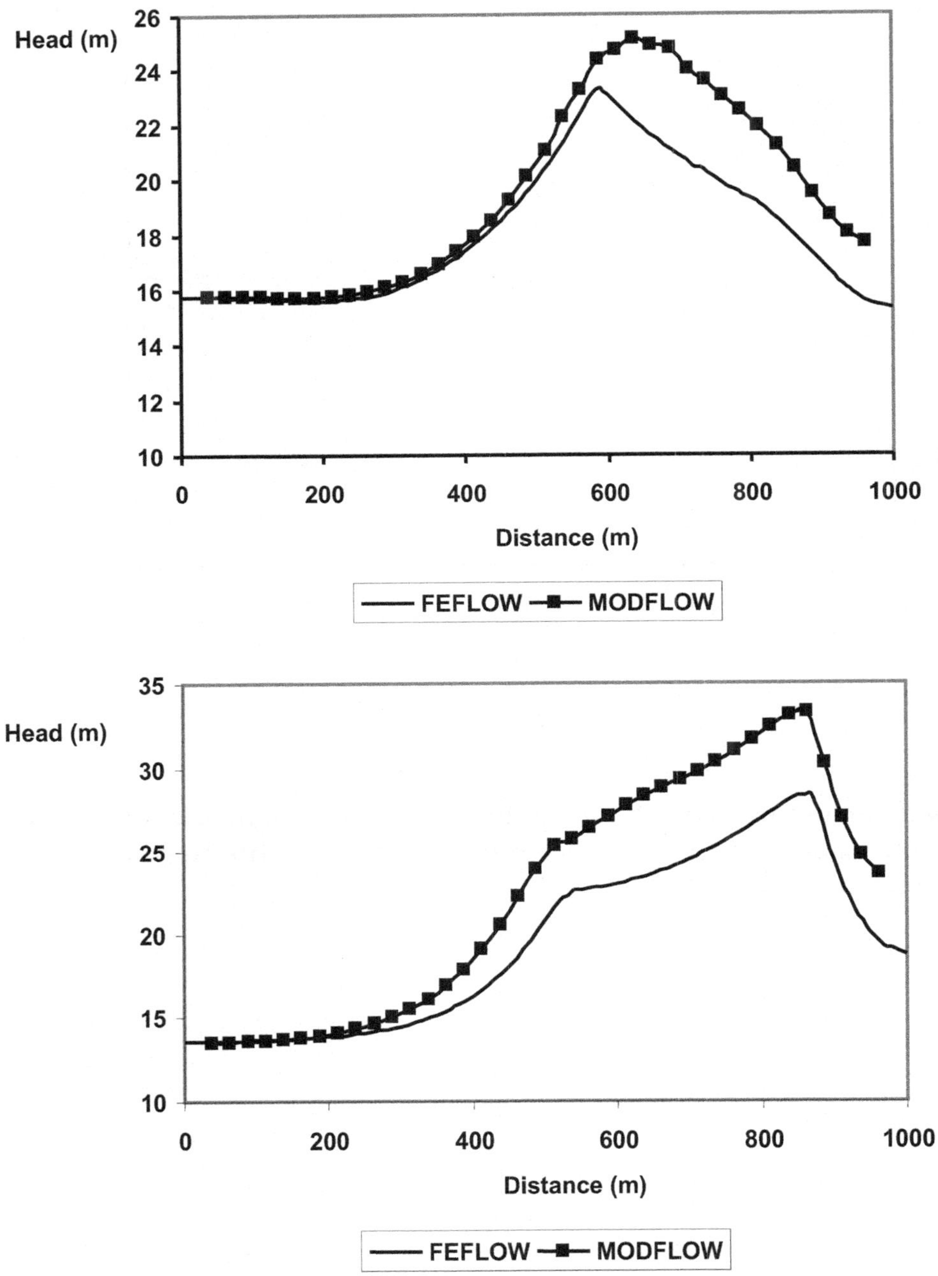

Figure 4.13 Comparison of the head distributions along a cross section drawn at y = 387.5 m for numerical Experiment 3. The FEFLOW® result is shown as a solid line, and the MODFLOW result is shown in the line with squares. (a) The head distributions at steady state, and (b) the head distributions at the end of the infiltration event.

(a)

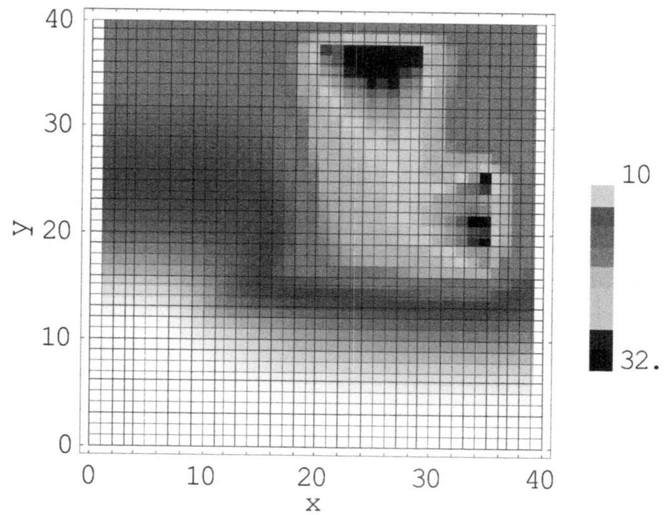

(b)

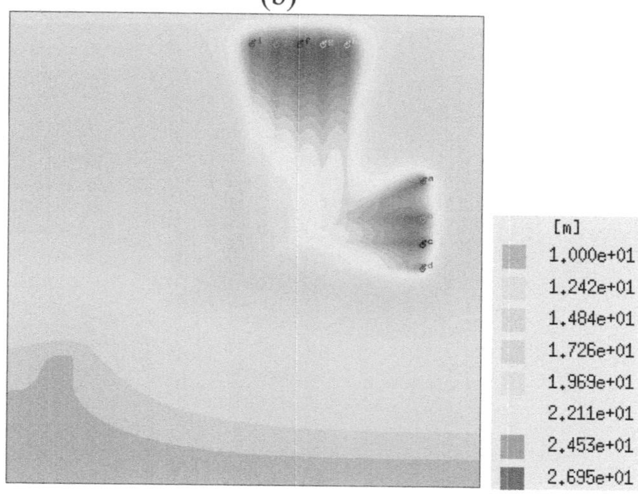

Figure 4.14 Color maps of the head distributions at t = 0.042 days in numerical Experiment 3. (a) MODFLOW and (b) FEFLOW®.

CHAPTER 5
DCM: A DUAL-CONDUCTIVITY PACKAGE FOR MODFLOW-2000

As discussed in Chapter 4, the single-continuum, smeared-conduit representation performs surprisingly well for confined conditions and simple networks. However, the single-continuum system does have limitations. The single-continuum approach becomes inaccurate if the conduit network becomes too dense, because the storage in the matrix is underestimated. For unconfined conditions, the approach becomes unwieldy and of questionable accuracy even for sparse conduit networks. Specifically, the transmissivity of a conduit depends on the saturated thickness, which depends in turn on the hydraulic head. Thus, the approach taken in Chapter 4 of calibrating the conduit transmissivity and storage properties may not work across the entire range of hydraulic heads for unconfined systems. Moreover, the single-continuum approach does not offer good control of the flow between conduits and background, an important calibration parameter.

The next level of complexity beyond the single-continuum model is the dual-conductivity model. In this approach, the diffuse background system is modeled as a continuum. Coupled to that continuum, and exchanging fluid with it, is a network of conduits. If the conduit network is dense, the conduit system can be modeled as a continuum, and the approach is identical to the dual-continuum model that has been used successfully in fractured rock studies. For mature karst aquifers, a relatively small number of conduits may dominate the flow. In this situation, the conduit system is a sparse network and not a continuum. Linear exchanges between the conduit system and the diffuse system are assumed, irrespective of whether the conduit system is modeled as a sparse network or a continuum. In this sense, the general approach is closely related to the classical dual-continuum model originally developed for fractured rock (Barenblatt et al., 1960).

The dual-conductivity model was implemented as a MODFLOW package called DCM as part of this project. The DCM package is currently limited to a single-layer aquifer. Specifically, the diffuse background is modeled as a single-layer aquifer, and the conduit system is assumed to be positioned within this single-layer aquifer. The approach incorporates significant three-dimensional information, despite the single-layer limitation. Specifically, the aquifer top and bottom and the top and bottom of the conduit system are input as functions of the *x-y* position. The remainder of this chapter describes the technical basis for the DCM package, validation tests, numerical experiments, and initial demonstration exercises.

TECHNICAL BASIS FOR THE DCM PACKAGE

Groundwater Flow Representation in MODFLOW

In MODFLOW, flow is conceptualized as occurring in multiple aquifer layers that may be stacked one upon the other. For a single layer using principal coordinates, the groundwater flow equations can be written

$$S(h)\frac{\partial h}{\partial t} = \frac{\partial}{\partial x}\left[T_x(h)\frac{\partial h}{\partial x}\right] + \frac{\partial}{\partial y}\left[T_y(h)\frac{\partial h}{\partial y}\right] + Q \qquad h \in \Omega \tag{5.1}$$

where h [L] is hydraulic head, T_x (T_y) [$L^{2/}T$] is transmissivity in the x (y) direction, S [-] is a storage term, and Q [L^3/T] is the volumetric source term per unit area of the aquifer. Equation (5.1) applies for both confined and unconfined aquifers with appropriate definitions of the head-dependent parameters T and S. Specifically, let $Z^{top}(x,y)$ denote the elevation of the top of the aquifer and $Z^{bot}(x,y)$ the bottom elevation. The x-direction transmissivity is then written

$$\begin{aligned} T_x &= 0 && h \le Z^{bot} \\ T_x &= K_x\left(h - Z^{bot}\right) && Z^{bot} < h < Z^{top} \\ T_x &= K_x\left(Z^{top} - Z^{bot}\right) && h \ge Z^{top} \end{aligned} \tag{5.2}$$

where K_x [L/T] is the x-direction hydraulic conductivity. The y-direction transmissivity is written similarly. For the storage term, the corresponding equation is

$$\begin{aligned} S &= 0 && h \le Z^{bot} \\ S &= S_Y && Z^{bot} < h < Z^{top} \\ S &= S_S\left(Z^{top} - Z^{bot}\right) && h \ge Z^{top} \end{aligned} \tag{5.3}$$

where S_Y [-] is the specific yield and S_S [L^{-1}] is the specific storage.

Dual-Conductivity Model

Equation (5.1) is to be solved over a specified region, Ω. For the dual-conductivity model, it is necessary to keep track of two hydraulic heads: one for the conduit and one for the diffuse background system. Hydraulic head in the background is defined over the entire region, Ω, as in the single conductivity case. The conduit hydraulic head is defined only for those spatial locations that correspond to a conduit. To be more specific, consider a system of n conduits, and let Ψ_i denote the spatial region occupied by the i-th conduit. Let the subscript c denote the conduit system, so that the hydraulic head of the conduit system becomes h_c. Similarly, let a subscript m denote the background system. The flow equations then become

$$S_m(h_m)\frac{\partial h_m}{\partial t} = \frac{\partial}{\partial x}\left[T_{mx}(h_m)\frac{\partial h_m}{\partial x}\right] + \frac{\partial}{\partial y}\left[T_{my}(h_m)\frac{\partial h_m}{\partial y}\right] + Q_m + \alpha(h_c,h_m)[h_c - h_m] \qquad h_m \in \Omega \tag{5.4a}$$

$$S_c(h_c)\frac{\partial h_c}{\partial t} = \frac{\partial}{\partial x}\left[T_{cx}(h_c)\frac{\partial h_c}{\partial x}\right] + \frac{\partial}{\partial y}\left[T_{cy}(h_c)\frac{\partial h_c}{\partial y}\right] + Q_c - \alpha(h_c,h_m)[h_c - h_m] \qquad h_c \in \Psi \tag{5.4b}$$

where Ψ is the entire region occupied by all conduits $\Psi = \Psi_1 \cup \Psi_2 \cup \Psi_3 \cdots \cup \Psi_n$. Note that while h_m is defined for the entire spatial region, Ω, h_c is defined only for region Ψ, which is a subset of Ω. If Ω and Ψ correspond, the system of Equation (5.4) is the dual-continuum model that is widely used in fractured rock modeling. For karst systems, the conduit network is generally not well represented as a continuum at the scale of interest, and the system of Equation (5.4) represents a sparse network of conduits coupled to a continuum background system. The transmissivity and storage terms in equation (5.4) are defined as in Equations (5.2) and (5.3), except that distinct top and bottom elevations may be used for the two flow systems: $Z_c^{top} \leq Z_m^{top}$ and $Z_c^{bot} \geq Z_m^{bot}$.

The final term in each equation in Equation (5.4) represents the movement of fluid between the two systems, with α quantifying the strength of the linear exchange. If α is 0, the conduit and background systems decouple, and flow in each system is independent. If the conduit is filled with water, α is a constant, independent of the head. If the conduit is only partially filled, this value may need to be decreased to account for the fact that only a fraction of the conduit surface area is available to transmit water. Thus, to model unconfined aquifers, α should be dependent on the hydraulic head. The situation is further complicated because flow can be either from the conduit to the background or vice versa, and the surface area available to transmit water depends on the flow direction. The simplest condition incorporating all of these constraints is the linear upwind condition

$$\alpha = \alpha_0 \left[\frac{\min\left(Z_c^{top}, \max\left(h_m, h_c, Z_c^{bot}\right)\right) - Z_c^{bot}}{Z_c^{top} - Z_c^{bot}} \right] \tag{5.5}$$

The term in brackets is unity if the conduit is completely filled with water and zero if both the conduit head and the diffuse head drop below the conduit base elevation. The parameter α_0 [T-1] is the linear exchange coefficient for a conduit filled with water. It is a property of the conduit and, in general, will be spatially variable. Theoretically, α_0 should be proportional to the product of the conduit surface area and background hydraulic conductivity. In practice, it is property of the system that is to be determined by calibration.

Turbulent Flow Model

Equation (5.4) presumes Darcy flow in the conduit system. To model turbulent flow in conduits, Equation (5.4) can be replaced with the following system:

$$S_m(h_m)\frac{\partial h_m}{\partial t} = \frac{\partial}{\partial x}\left[T_{mx}(h_m)\frac{\partial h_m}{\partial x}\right] + \frac{\partial}{\partial y}\left[T_{my}(h_m)\frac{\partial h_m}{\partial y}\right] + Q_m + \alpha(h_c, h_m)[h_c - h_m] \qquad h_m \in \Omega$$

$$S_c(h_c)\frac{\partial h_c}{\partial t} = -\frac{\partial}{\partial x}[q_{cx}] - \frac{\partial}{\partial y}[q_{cy}] + Q_c - \alpha(h_c, h_m)[h_c - h_m] \qquad h_c \in \Psi \tag{5.6}$$

where

$$q_{cx} = -T_{cx}^{1}(h_c)\left|\frac{\partial h_c}{\partial x}\right|^{-1/2}\frac{\partial h_c}{\partial x} \tag{5.7}$$

in the situation of turbulent flow, and

$$q_{cx} = -T_{cx}(h_c)\frac{\partial h_c}{\partial x} \tag{5.8}$$

in the situation of Darcy (laminar) flow, with analogous expressions for the y-components of flux. The parameter T^1_{cx} is a proportionality constant incorporating geometrical properties of the conduit and friction factor. It is regarded here as a calibration parameter similar to aquifer transmissivity.

The current version of DCM allows the user to select either Darcy flow (5.8) or turbulent flow (5.7) in the conduit. Transition between laminar and turbulent flow occurs at Reynolds numbers of about 2200, and both are relevant to flow in karst conduits. A revised version of DCM that selects the flow law dynamically based on the local flow conditions will be released in the near future.

IMPLEMENTATION OF THE DCM PACKAGE

The dual-conductivity model was implemented as a MODFLOW-2000 package called the DCM package. The implementation is based on the LPF internal flow package and provides an alternative to the LPF. Two MODFLOW layers are used in the DCM package. Layer 1 is the conduit layer, and Layer 2 is the diffuse layer. The conduit layer is assumed to be completely contained by the diffuse layer. In the conduit layer, only those cells crossed by a conduit are active. Thus, in many simulations, the majority of the conduit cells will be designated inactive. Specific modifications to MODFLOW required for the DCM package are

1. Error checking for layer overlap was bypassed. The LPF flow package checks to see that layers do not overlap. This step was bypassed in the DCM package because the conduit layer should be completely contained by the diffuse layer.

2. The vertical leakance calculation was replaced by the exchange model described by Equation (5.5). In the current single-layer DCM package, there is no vertical leakage, but there is diffuse-conduit exchange. The diffuse-conduit exchange model simply replaces the vertical leakance calculation.

3. A new algorithm for drying-rewetting was implemented. As discussed in Chapter 3, small changes in hydraulic head can lead to relatively large changes in saturated thicknesses (and, hence, transmissivity) for conduits with comparatively small vertical extent. This sensitivity to small changes in hydraulic head tends to amplify the well-known drying-rewetting instability in MODFLOW. In the new drying-rewetting algorithm, if hydraulic head drops below the bottom elevation of a cell, the calculated hydraulic head is simply set equal to the cell bottom elevation. In contrast to the LPF package, the cell is not rendered inactive but continues without any special treatment with the new hydraulic head. To prevent flow out of a dry cell, interblock conductances are calculated using upstream weighting of the saturated thickness instead of the standard arithmetic average used in the LPF package. Preliminary tests using this new algorithm suggest improved stability relative to the LPF drying-rewetting algorithm. To make full use of this algorithm, adaptive timestepping should also be used; however, this capability has not been implemented yet.

4. It should be noted that the DCM package is currently limited to single-layer aquifers. Thus, the package will model the situation shown in Figure 5.1(a), but not the multilevel configuration shown in Figure 5.1(b). To model the configuration shown in Figure 5.1(b) or an aquifer with multiple layers with disparate properties (Figure 5.1(c)), a 3D version of the DCM package will be required.

VALIDATION TESTS

Test 1: Dual-Conductivity Confined System

This simulation is designed to test the DCM representation of dual-conductivity flow in a confined system. The configuration is shown in Figure 5.2. The system is initially in steady state with hydraulic head of 0 meters. The Darcy flow law is used. At t = 0, the head in the conduit system on the left boundary is increased from 0 to 5 meters. The simulation then proceeds for 100 minutes. Inputs parameters are listed in Table 5.1. The system has translational symmetry in the *y*-direction and can be modeled as a one-dimensional system. However, it was implemented as a two-dimensional system, and symmetry in the *y*-direction was checked as part of the validation. In addition, the results of the numerical simulation were compared with an independent numerical solution.

Table 5.1
Parameters for the dual-conductivity validation test

Parameter	Value	Units
Conduit Transmissivity	1	m^2/min
Diffuse Transmissivity	1	m^2/min
Conduit-Diffuse Exchange Term	10^{-4}	min^{-1}
Conduit Storativity	0.001	dimensionless
Diffuse Storativity	0.001	dimensionless

For a one-dimensional confined system and presuming Darcy flow, Equations (5.6) become

$$S_m \frac{\partial h_m}{\partial t} = \frac{\partial}{\partial x}\left[T_{mx} \frac{\partial h_m}{\partial x}\right] + \alpha_0 \left[h_c - h_m\right] \tag{5.9a}$$

$$S_c \frac{\partial h_c}{\partial t} = \frac{\partial}{\partial x}\left[T_{cx} \frac{\partial h_c}{\partial x}\right] - \alpha_0 \left[h_c - h_m\right] \tag{5.9b}$$

where the *S, T*, and α_0 parameters are all independent of head. These equations were solved using the *NDSolve* routine of the commercial software Mathematica™. The *NDSolve* routine uses sophisticated adaptive algorithms for one-dimensional systems of partial differential equations and the resulting solutions are highly accurate.

Hydraulic heads calculated by DCM are compared to the benchmark solutions at $t = 5$ minutes and $t = 20$ minutes in Figure 5.3. The benchmark solutions are shown as curves and the DCM results as individual data points. The agreement is excellent. The maximum absolute difference between the simulated and benchmark hydraulic head in the conduit is less than 5 cm, or less than 1% of the 5-meter head drop in the system. The maximum deviation in the diffuse head is even smaller.

Test 2: Unconfined System

These simulations are designed to test the DCM representation of flow in an unconfined system. The configuration is the same as in Figure 5.2. The base of the conduit and the diffuse system are specified at 0 meters elevation, while the top of the conduit is at 3 meters, and the top of the diffuse system at 10 meters. The system is initially at steady state, with hydraulic head of 0 meters. At $t = 0$, the head in the conduit system on the left boundary is increased to 5 meters. Inputs parameters are listed in Table 5.2. Two variants are considered, depending on the value of the exchange parameter.

Table 5.2
Parameters for the unconfined flow validation test (Test 2)

Parameter	Value	Units
Conduit Transmissivity	1	m^2/min
Diffuse Transmissivity	1	m^2/min
Conduit-Diffuse Exchange Term	0 or 10^{-4}	min^{-1}
Conduit Storativity	0.001	dimensionless
Diffuse Storativity	0.001	dimensionless
Conduit Specific Yield	0.1	dimensionless
Conduit Specific Yield	0.1	dimensionless

In the first variant, Test 2, the exchange term is set to zero and the simulation proceeds for 120 minutes. With no exchange, the conduit head is governed by the standard single-conductivity model, and the results of the DCM simulation can be compared with the results of the standard MODFLOW LPF package. It should be recognized, however, that DCM and LPF use different techniques for averaging the saturated thickness between two grid cells. Specifically, LPF uses simple arithmetic averaging of the saturated thickness and DCM uses the upstream weighting method. Thus, DCM and LPF are not expected to give the same result. The validation test is designed to provide some information on the consequences of the new averaging scheme in addition to verifying that the equations are implemented correctly.

Results of Test 2 are shown in Figure 5.4 for $t = 1$, 10, and 60 minutes. The agreement is good. The small deviations near the leading edge of the water pulse are due to the different averaging schemes and are to be expected.

In the second variant, Test 2a, the exchange term is set to 10^{-4} min^{-1} and the simulation proceeds until steady state is reached. Once steady state is reached, the calculated hydraulic heads can be used to verify on a cell-by-cell basis that the water balance is maintained. Steady-state results are shown in Figure 5.5. Hand calculations verify that the water balance is correct. For example, for a cell located about 90 meters from the outlet, the water balance is correct to one part in 10^5.

Test 3: Turbulent Flow

This simulation is designed to test the DCM representation of turbulent flow. The configuration is similar to that of Figure 5.1. The system is initially at steady state, with hydraulic head of 0 meters. At t = 0, the head in the conduit system on the left boundary is increased to 5 meters. The system is confined and there is no coupling between the conduit and diffuse system in this test. Inputs parameters are listed in Table 5.3.

Table 5.3
Parameters for the turbulent flow validation test (Test 3).

Parameter	Value	Units
Conduit Transmissivity at Unit Head Gradient	0.25	m^2/min
Diffuse Transmissivity	1	m^2/min
Conduit-Diffuse Exchange Term	0	min^{-1}
Conduit Storativity	0.001	dimensionless
Diffuse Storativity	0.001	dimensionless

In 1D with no conduit/diffuse interaction, the equation governing evolution of the hydraulic head in the conduit is

$$S_c \frac{\partial h_c}{\partial t} = \frac{\partial}{\partial x}\left[T_c^1 \left| \frac{\partial h_c}{\partial x} \right|^{-1/2} \frac{\partial h_c}{\partial x} \right] \tag{5.10}$$

where T_c^1 is the effective transmissivity at unit gradient and all other parameters have been defined previously. This equation was solved using the *NDSolve* routine of the commercial software Mathematica™, as in Test 1.

Results are shown in Figure 5.6 for $t = 5$, 10 and 20 minutes. The maximum error is about 17 cm, which is 3.5% of the 5 meter head drop. The error is smaller in regions where the grid spacing is smallest, which suggests that the error is due to the limited resolution of the grid. To test this hypothesis, another simulation with twice as many computational cells (half the grid spacing) was constructed. In this new simulation, the maximum error was about 8 cm or about 1.5% of the total head drop. This decrease in error with increasing grid resolution suggests that the small errors in Figure 5.6 are caused by the finite resolution of the numerical grid.

NUMERICAL EXPERIMENTS USING THE DCM PACKAGE

Sensitivity to the exchange coefficient α_0 is illustrated in Figure 5.7. This example is identical to the single-conduit configuration of Figure 4.3, but is modeled with MODFLOW-2000 using the prototype DCM package. Spring hydrographs (discharge versus time) are shown for three different values of the coupling coefficient α_0. Recall the recharge event is a square pulse with a duration of 2 hours (0.08 days) and magnitude of 0.5 m^3/s (43,200 m^3/day). For small values of α_0, the interaction between the conduit and the diffuse system is relatively weak, and the spring discharge responds rapidly to the recharge event. As α_0 increases, more water is transferred to the diffuse system, and the spring hydrograph responds more slowly and has a slower recession. This sensitivity to the exchange coefficient indicates that it is an important parameter for use in calibration. Numerical tests indicate the hydrograph becomes nearly independent of conduit storage if α_0 is sufficiently large.

A more complicated example is summarized in Figure 5.8. The configuration is the same as shown as in Figure 4.11. Hydraulic heads along the transect y = 387.5 m are shown for two times: at steady state and after 2 hours of recharge into the conduit system. Also shown is spring discharge versus time. Results for a single-continuum MODLFOW simulation are shown for comparison. The MODFLOW-DCM simulation reproduces the FEFLOW® simulation much better than the single-continuum MODFLOW. This better performance is due in part to the additional calibration parameter α_0, which is not available in a single continuum run, and in part, to the fact that the dual-conductivity simulation does not artificially reduce the diffuse storage as does the single conductivity simulation.

The final example is a single conduit in an unconfined aquifer. The configuration is similar to that of the single-conduit example of Figure 4.3, except the conduit is in this case aligned with the grid. Cross-sectional and plan views are shown in Figures 5.9(a) and 5.9(b). Results are provided in Figure 5.10. The water level is below the conduit bottom in the upper reaches of the conduit. This situation was deliberately selected as a demanding test of the new drying–rewetting algorithm. Without this new algorithm, MODFLOW failed to converge for these conditions.

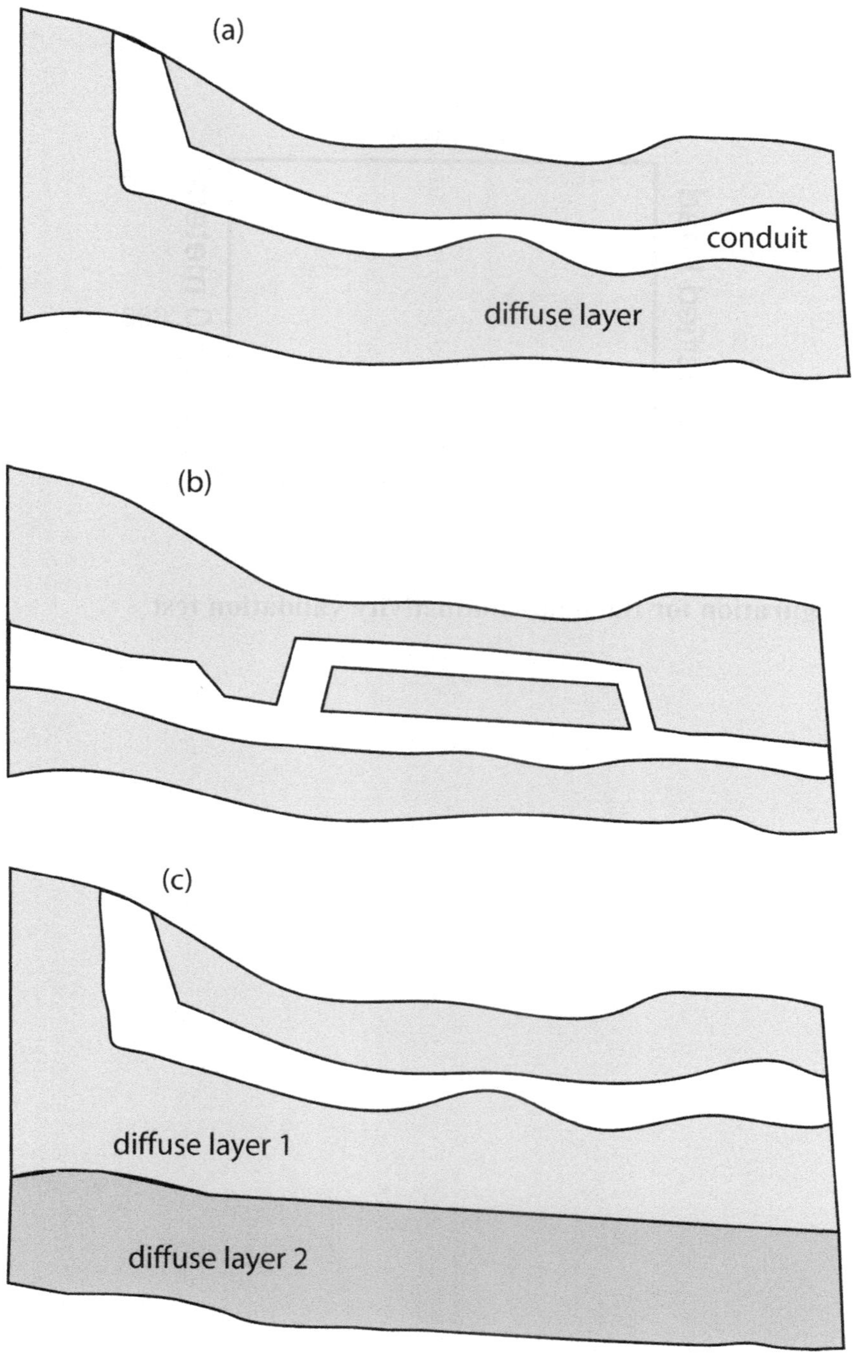

Figure 5.1 Hypothetical cross sections illustrating the range of applicability of the DCM module for MODFLOW. The DCM module will model a single-layer aquifer containing a single-level conduit system, as in (a). The DCM module will not model a multilevel conduit system like that shown in (b), or a multiple layer aquifer similar to the one shown in (c).

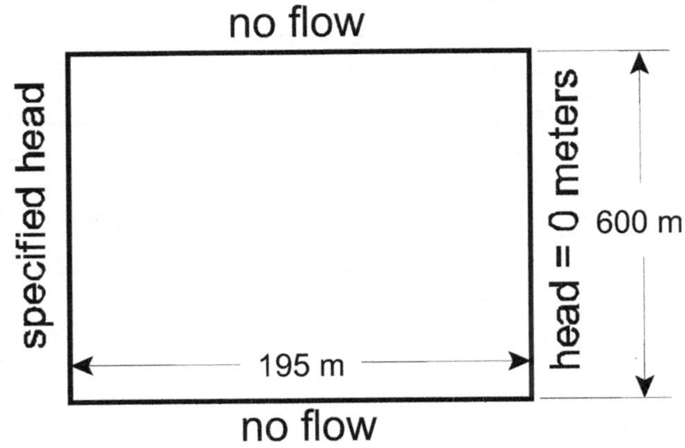

Figure 5.2 Configuration for the dual-conductivity validation test

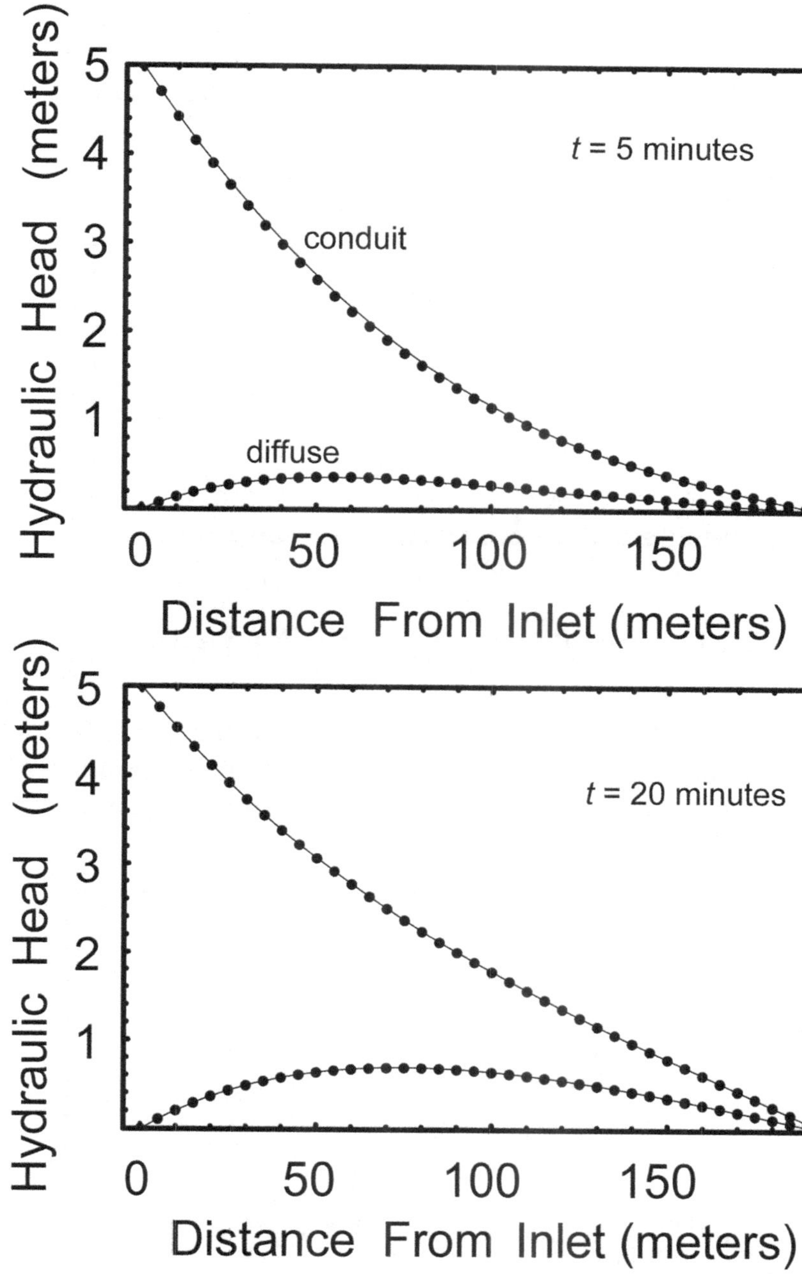

Figure 5.3 Results of the dual-conductivity validation test at *t* = 5 minutes and *t* = 20 minutes. The benchmark results are shown as curves and the DCM results as individual data points. In each figure, the upper set represents hydraulic head in the conduit system and the lower set represents the hydraulic head in the matrix system.

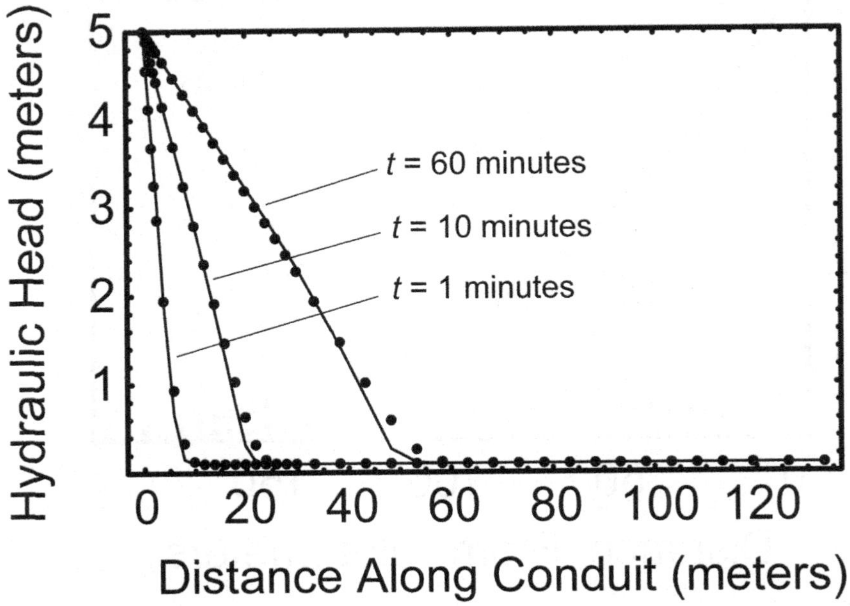

Figure 5.4 Results of the unconfined flow validation test (Test 2) at $t = 1$, 10, and 60 minutes. The hydraulic heads in the conduit as calculated by the standard MODFLOW LPF package are shown as curves and the DCM results as individual data points.

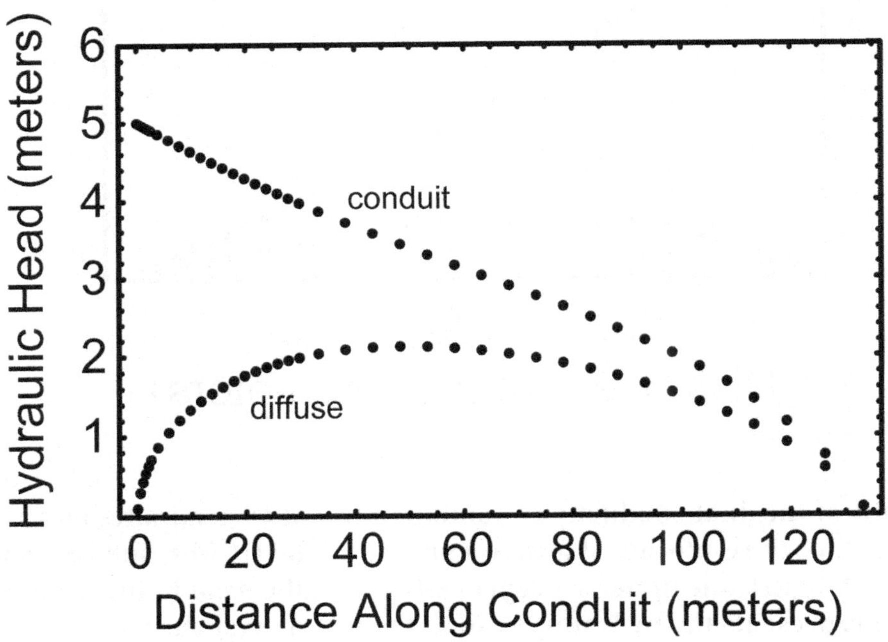

Figure 5.5 Steady-state results for the unconfined flow validation test (Test 2a). This simulation was used to verify that water balance is properly maintained in the DCM package.

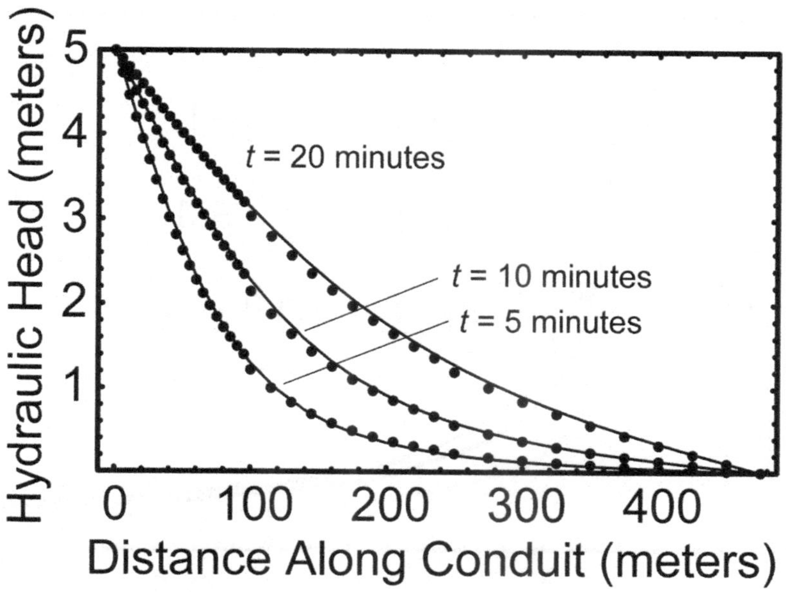

Figure 5.6 Results of the turbulent flow validation test (Test 3) at t = 5, 10, and 20 minutes. The benchmark hydraulic heads in the conduit are shown as curves and the DCM results as individual data points.

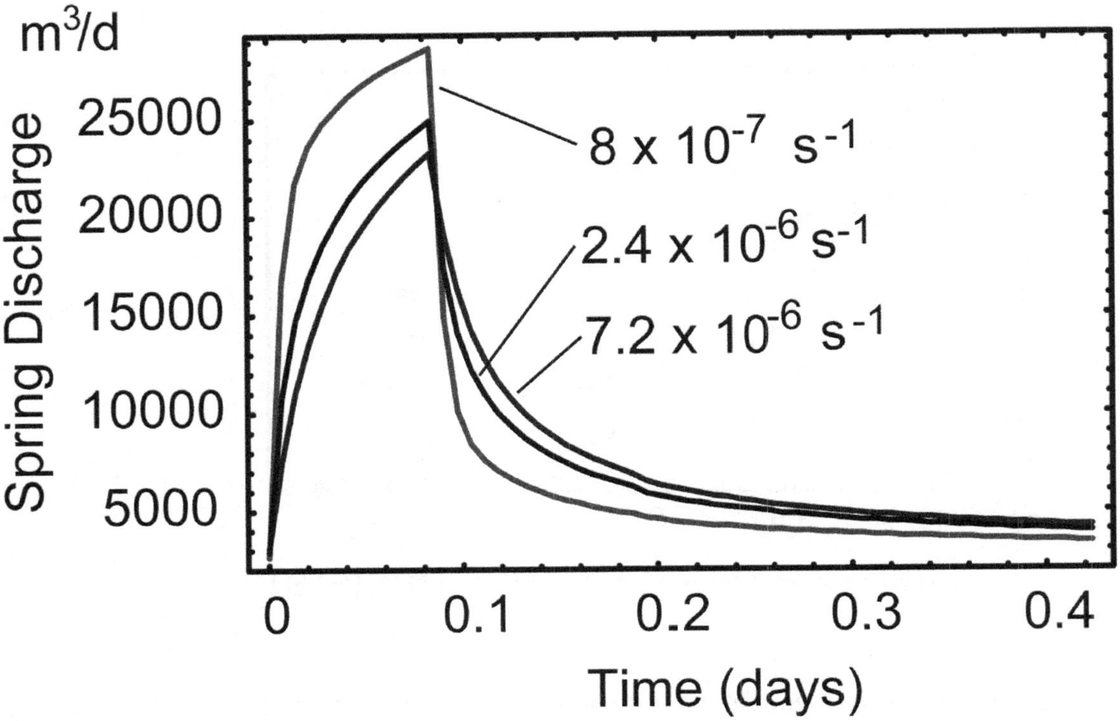

Figure 5.7 Sensitivity of spring discharge to the diffuse-conduit interaction parameter α_0. The flow configuration is the single-conduit system shown in Figure 4.3.

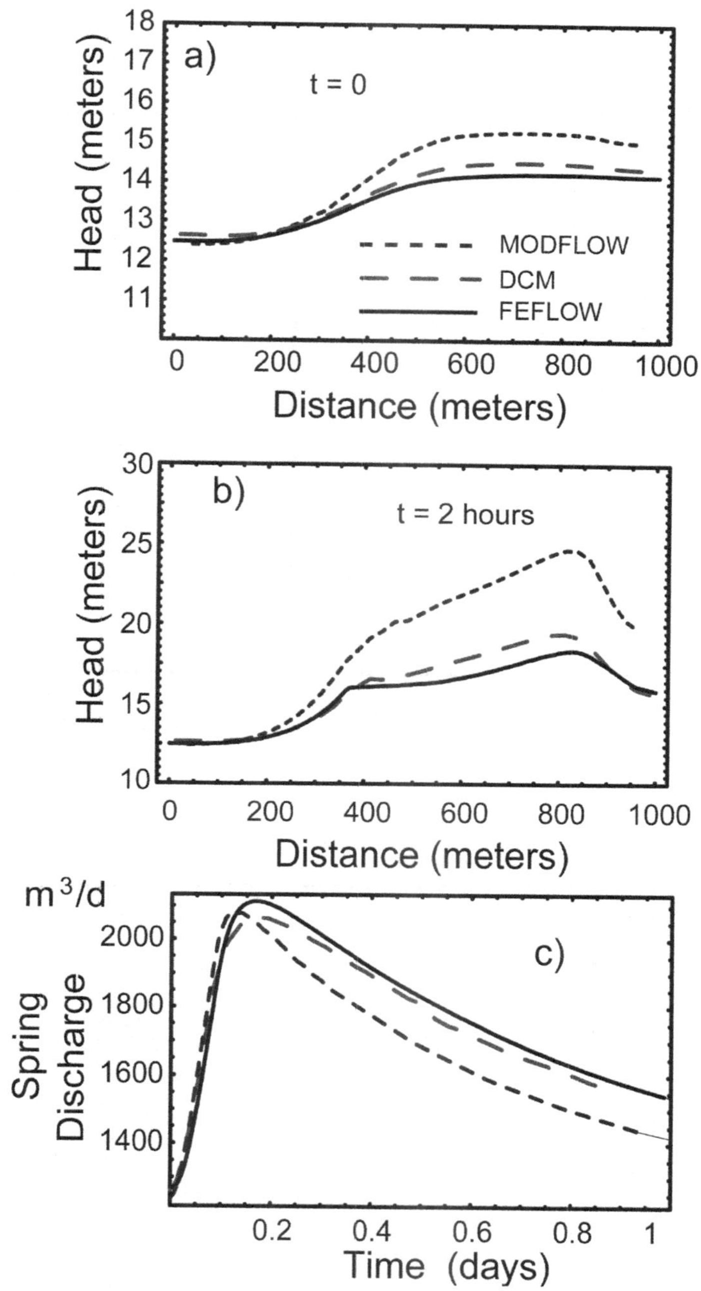

Figure 5.8 Hydraulic head and spring discharge as calculated by MODFLOW-DCM and FEFLOW® for the multiple conduit example of Figure 4.11. Shown are the (a) steady-state hydraulic head along the scan line y = 387.5 m (see Figure 4.11), (b) hydraulic head along the same scan line after a 2-hour infiltration event, and (c) spring discharge versus time.

(a)

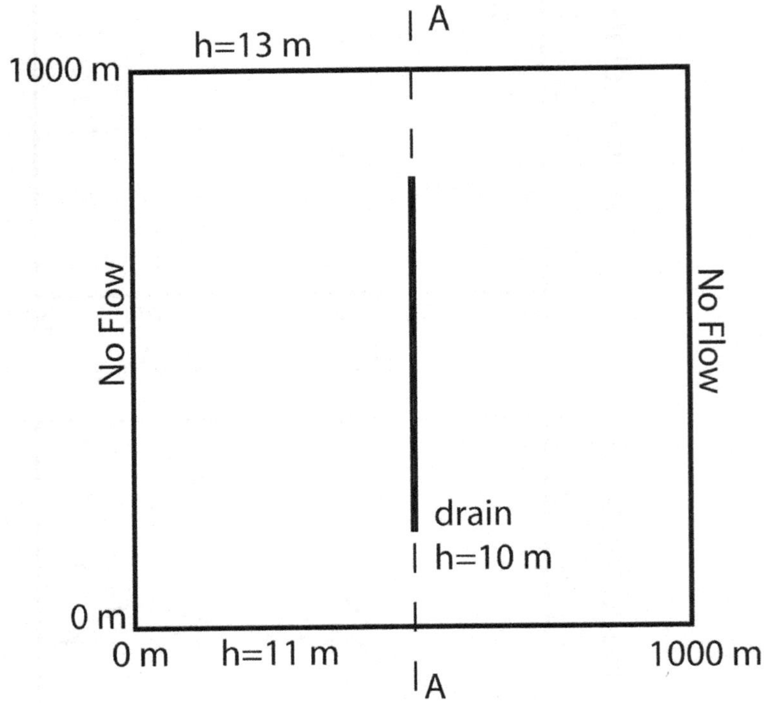

(b)

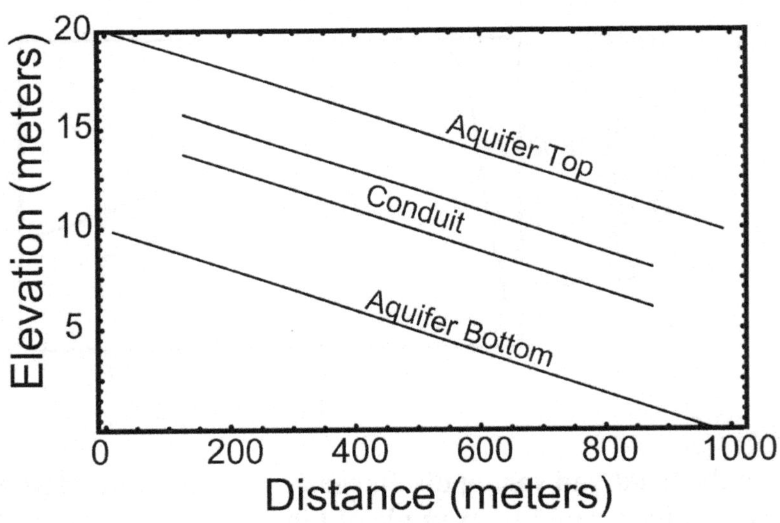

Figure 5.9 Hypothetical unconfined aquifer used as a demonstration of MODFLOW-DCM. The conduit configuration is shown in (a) map view and (b) cross section.

(a)

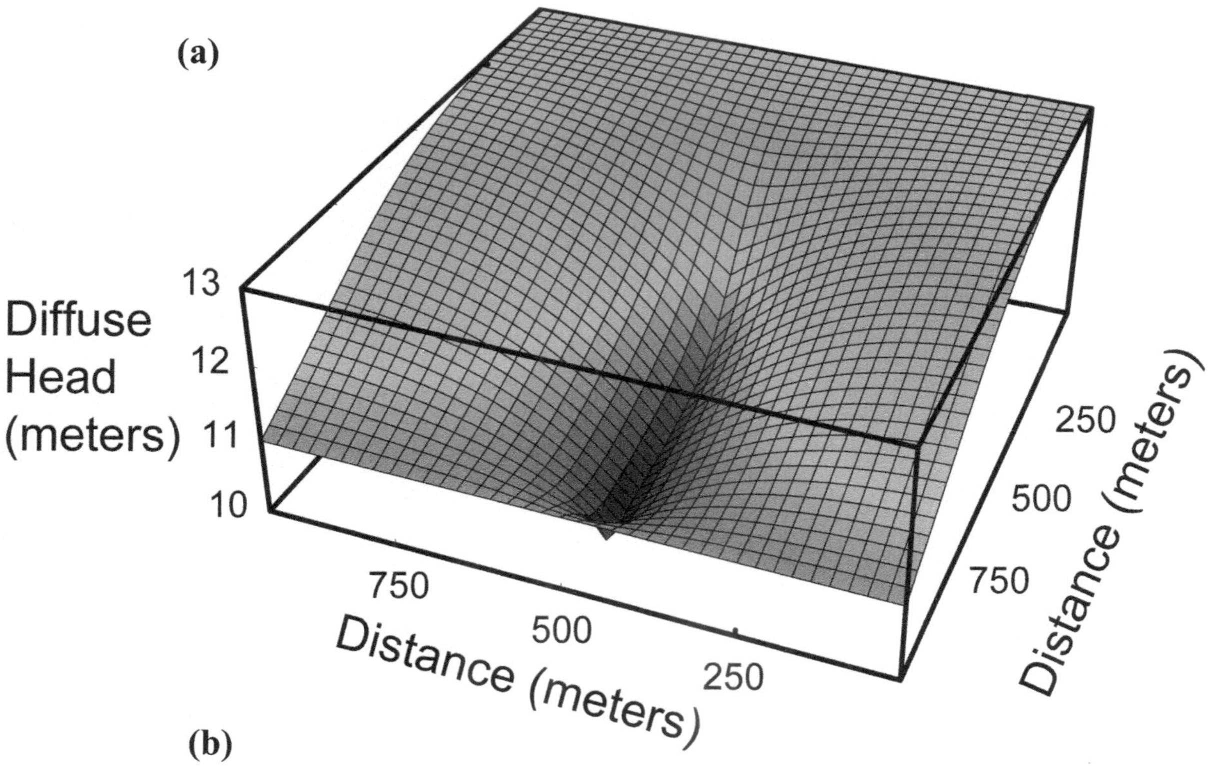

(b)

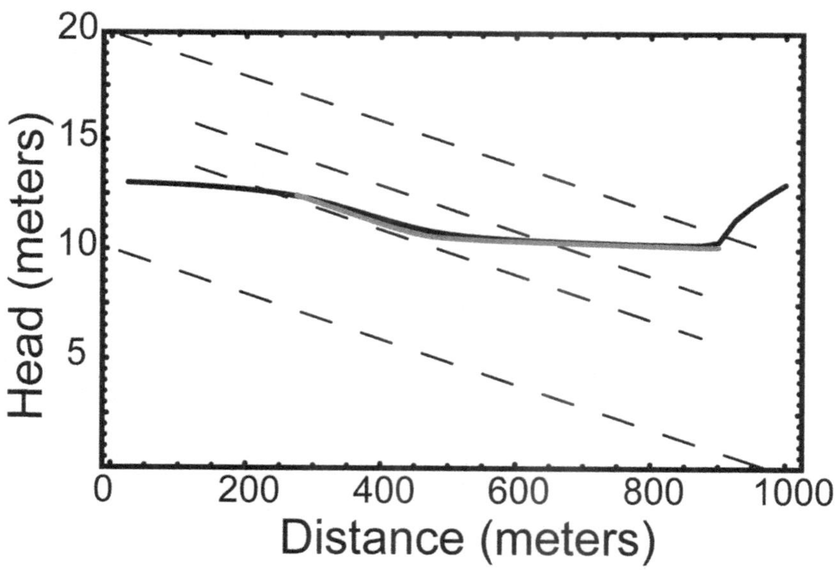

Figure 5.10 Hydraulic head in steady state for the unconfined aquifer of Figure 5.4, as calculated by MODFLOW-DCM. The hydraulic head is shown as a surface plot in (a). The hydraulic head along the scan line A-A (illustrated in Figure 5.9) is shown in (b).

CHAPTER 6
DCM MODEL OF THE BARTON SPRINGS SEGMENT OF THE EDWARDS AQUIFER

As an initial trial of the DCM package for MODFLOW-2000, the DCM software was used to model the Barton Springs segment of the Edwards Aquifer. The Barton Springs Edwards Aquifer Conservation District (BSEACD) has constructed a MODFLOW-96 model of the Barton Springs segment (Scanlon et al., 2001, 2003). In this task of the project, the BSEACD model for Barton Springs was converted to a MODFLOW–DCM model as an initial trial of the DCM approach. The objectives were to test the stability and numerical performance of the DCM code, to identify any shortcomings, and to gain some experience with applying an explicit conduit code to a real-world aquifer. A fully calibrated DCM model of Barton Springs will be developed in the future and was not attempted for this initial study.

STUDY AREA AND HYDROGEOLOGICAL SETTING

The Barton Springs segment is a sub-section of the Edwards Aquifer located in south-central Texas (Figure 6.1). The Edwards Aquifer is a Cretaceous age limestone aquifer which exhibits well-developed karstic features—such as caves, sinkholes, and swallets—and associated rapid karst flow. Rapid groundwater velocities, indicative of karst aquifers, have been measured in the Edwards Aquifer using results from dye tracer tests (Hauwert et al., 2002).

The Edwards Aquifer is divided into three sub-sections: the San Antonio segment, the Barton Springs segment, and the Northern segment. The Barton Springs segment is terminated by a groundwater divide near Kyle on the west end and the Barton Springs along the Colorado River at the eastern down-gradient end. In general, the Edwards Aquifer is structurally controlled by the uplift of the Edwards Plateau along the Balcones Fault Zone. This faulting has developed three zones associated with the Edwards Aquifer: the contributing zone, the recharge zone, and the confined zone. Normal faulting along en echelon faults and grabens associated with the uplift of the Edwards Plateau have resulted in exposure of the underlying Glen Rose Formation to the north and northwest of the Edwards Aquifer. This region is referred to as the contributing zone to the Edwards Aquifer. Faulting to the south and southeast has exposed the Edwards Aquifer at the surface. This region is referred to as the recharge zone. Farther to the south and southeast, younger, less-permeable units overlay the Edwards Aquifer causing the aquifer to be confined. The southern and southeastern extent of the Edwards Aquifer is defined by the transition from fresh to saline water, commonly designated as water with total dissolved solids in excess of 1,000 ppm.

The Barton Springs segment, including the contributing zone, is approximately 40 km long and 20 km wide. The recharge zone of the Barton Springs segment covers 250 km^2. Along this segment, the Edwards Aquifer has an approximate thickness of 30-180 m, with thickness greatest on the down-dip side. Most fault orientations are coincident with the overall Balcones Fault Zone orientation, which tends to be northeast-southwest in the Barton Springs segment. Also oriented coincident with the Balcones Fault zone orientation are the main groundwater karst conduits whose locations and orientations have been surmised using results from tracer tests (Hauwert et al., 2002). Tracer test results have also allowed identification of three principal subbasins within the Barton Springs segment: Cold Springs groundwater basin, Sunset Valley groundwater basin, and Manchaca groundwater basin.

THE BSEACD MODFLOW MODEL

The BSEACD model has a single layer with 7,043 active cells and uses the confined-unconfined mode of MODFLOW. The grid cells are 1,000 feet long and 500 feet wide. The grid is aligned with the predominant faults in the Edwards formation and is thus rotated 45° from the east–west direction (see Figure 6.1). The model extends from a presumed groundwater divide near Buda, Texas, in the south to the Colorado River near downtown Austin along the northern boundary. The eastern boundary corresponds to the fresh-saline water interface, and the western boundary corresponds to the Mount Bonnell fault.

All boundaries were modeled as no-flow boundaries. Barton Springs and Cold Springs were modeled as drains with very high drain conductances. Recharge had two components. Eighty-five percent of the recharge was presumed to be concentrated along losing streams, and the remaining 15% was distributed uniformly across the recharge zone. The aquifer was divided into nine zones for the purpose of assigning transmissivity. Higher transmissivities were assigned to zones aligned with the major flow paths as inferred from tracer tests.

BSEACD constructed a steady-state model and a transient model. The steady-state model used recharge and pumping data from 1992. Transmissivities in the nine zones were calibrated to match hydraulic head from this period. The transient model used monthly recharge and pumping for the 10-year period from 1989 through 1998.

The BSEACD model was constructed to model spring flow during drought periods and aquifer-wide water level declines due to future increased pumping. From this perspective, the model was successful because it generally reproduced spring flow histories during the 1989 to 1998 period. However, the model does have some limitations. Simulated recessions in the spring flows tend to be a little slower than observed. Simulated spring flows during high-flow periods tend to be larger than observed. Water-level hydrographs in monitoring wells appear to fluctuate more strongly in response to changes in recharge than the observed hydrographs. Simulated heads in the southwest fringe of the model tend to be too high. Most importantly, simulated spring flows during extreme droughts tend to be too large. To address that problem, BSEACD has developed an alternative calibration to simulate extreme low-flow periods.[2]

The limitations of the current BSEACD model may reflect modeling assumptions or limitations of the input data. Another possibility is that these limitations are reflecting the inherent limitations of using a single-continuum model with a Darcy flow law to model a dynamic karst system. One of the multiyear objectives of the karst modeling project is to test the capability of the dual-conductivity model to model dynamic karst systems. As an initial step in that direction, the BSEACD model was converted to a dual-conductivity model described in the next section.

MODFLOW/DCM MODEL

The BSEACD model was converted to a MODFLOW/DCM model by (1) adding a conduit layer, (2) reducing transmissivity in the high transmissivity zones to levels more typical for a diffuse flow system, and (3) increasing specific storage and specific yield in the diffuse

---

[2] Brian Smith, personal communications

system to values more typical for a diffuse flow system. Modifications (2) and (3) are necessary because the conduit system is now accounting for the majority of flow and all of the fast flow.

The conduit system is shown in Figure 6.2. Conduits were placed underneath each of the losing streams. This configuration allows direct recharge into the conduit system. Each of these branch conduits was then connected to the two main conduits that terminate at Barton Springs. The positions of these main conduits are based on results of tracer tests (Hauwert et al., 2002) and observed groundwater troughs. The positions are approximate, and will be refined in future studies, but are considered adequate for testing purposes. Conduits that terminate at Cold Springs were ignored in this initial test. The main conduits were assigned an effective conductivity value of 2200 feet/day (7.76×10^{-3} m/s). The branch conduits were assigned 200 feet/day (7.05×10^{-4} m/s). These values were selected based on initial simulations. The exchange parameters for the conduits were assigned values in the range 0.001–0.01 day^{-1} (1.16×10^{-8}-$1.16 \times 10^{-7}\ \text{s}^{-1}$).

Other aspects of the model were left unchanged from the BSEACD model. The model uses the recharge, horizontal flow barrier, drain, and well packages.

SIMULATION RESULTS

Steady-State Model

Figure 6.3 shows calculated hydraulic heads in the conduit and diffuse systems as color density plots. Figure 6.4 shows the diffuse system head as a surface. Positions of the conduits are marked on this plot in black. Hydraulic head increases from 430 feet (131 m) at Barton Springs to more than 800 feet (244 m) in the western edge of the recharge zone. Hydraulic gradients are generally low in the confined region due to the influence of the major conduits. Hydraulic gradients are larger in the recharge zone. These features are in general agreement with hydraulic head measured during the 1992 period. Contours of observed head (Scanlon et al., 2001, 2003) from this period are shown for comparison in Figure 6.5.

Transient Model

A 1-year transient simulation was also performed. Hydraulic heads calculated with the steady-state model were used as the initial condition. The simulation used 1-month recharge periods. The recharge values are shown in Figure 6.6. These are intended to approximate the recharge for 1989. The time steps ranged from 1–3 days, depending on the recharge period. Pumping was constant during the simulation and the same as in the steady-state model.

Color density plots of the calculated hydraulic heads in the conduit and diffuse system at the end of month 5 are shown in Figure 6.7. Month 5 is a period of high recharge. The conduit head is larger than the diffuse system head during this period, and the flow is from the conduit to the diffuse system. Groundwater ridges develop in the diffuse system in the recharge zone and to a lesser extent in the confined zone.

Figure 6.8 shows similar color density plots at the end of the 1-year simulation. The last 5 months of the simulation have no recharge and the hydraulic heads at the end of the simulation period are considerably lower than in Figure 6.7. Moreover, significant groundwater troughs have developed along the conduit positions.

The spring-flow hydrograph and a water-level hydrograph for two wells are shown in Figure 6.9. The spring flow responds very quickly to changes in recharge. For the transient simulation, the springs respond within hours of a change in recharge. This response time can be adjusted by changing specific yield in the conduit system or the conduit/diffuse exchange parameter. The water-level hydrograph for a location near Barton Springs is very unresponsive (subdued) and changes by less than 2 feet during the 1-year simulation. At the location near the southernmost conduit the water level responds slower than the springs and fluctuates by about 10 feet during the simulation.

The general responses of the hydrographs are roughly consistent with observed hydrographs. Of particular interest is the relatively subdued response for water level near the conduit head. In the BSEACD single-continuum MODFLOW simulation, the water level in that location is overly responsive to changes in recharge and changes by about 50 feet during the 1989 period. The less responsive behavior in the DCM simulations is encouraging and suggests that the additional flexibility inherent in the DCM approach may allow better representation of dynamic behavior in karst aquifers.

Numerical Performance

The main goal of the Barton Springs simulations was to test the numerical stability and performance of the DCM package in a full-scale application, and to identify any needed refinements. Based on the experience with the Barton Springs Segment and with the numerical experiments described here, the following observations can be made regarding numerical performance.

1. For confined systems, no problems were observed with either the laminar or turbulent options.

2. For unconfined systems, iteration is required at each time step. When the water level in a conduit approaches the conduit bottom elevation, the iterations often fail to converge.

3. For transient unconfined simulations, converged solutions were obtained in all cases by introducing a damping factor in the iteration and by limiting the size of the time step.

4. Steady-state solutions were obtained in all cases by using transient simulations to reach steady-state conditions.

5. Despite the need to limit the size of the time steps, the Barton Springs 1-year transient simulation required only 10 minutes to execute on a 900 megahertz SUN UltraSparc III computer.

Time-step failure due to drying/rewetting is an intrinsic problem in the MODFLOW iteration approach, but is amplified by the highly transmissive conduits. Given that converged solutions were obtained in all cases by limiting the magnitude of time steps, an obvious next improvement would be to introduce an adaptive time step. Specifically, MODFLOW should be modified to reduce the time step and retry the step following a time-step failure. This modification is planned for future versions.

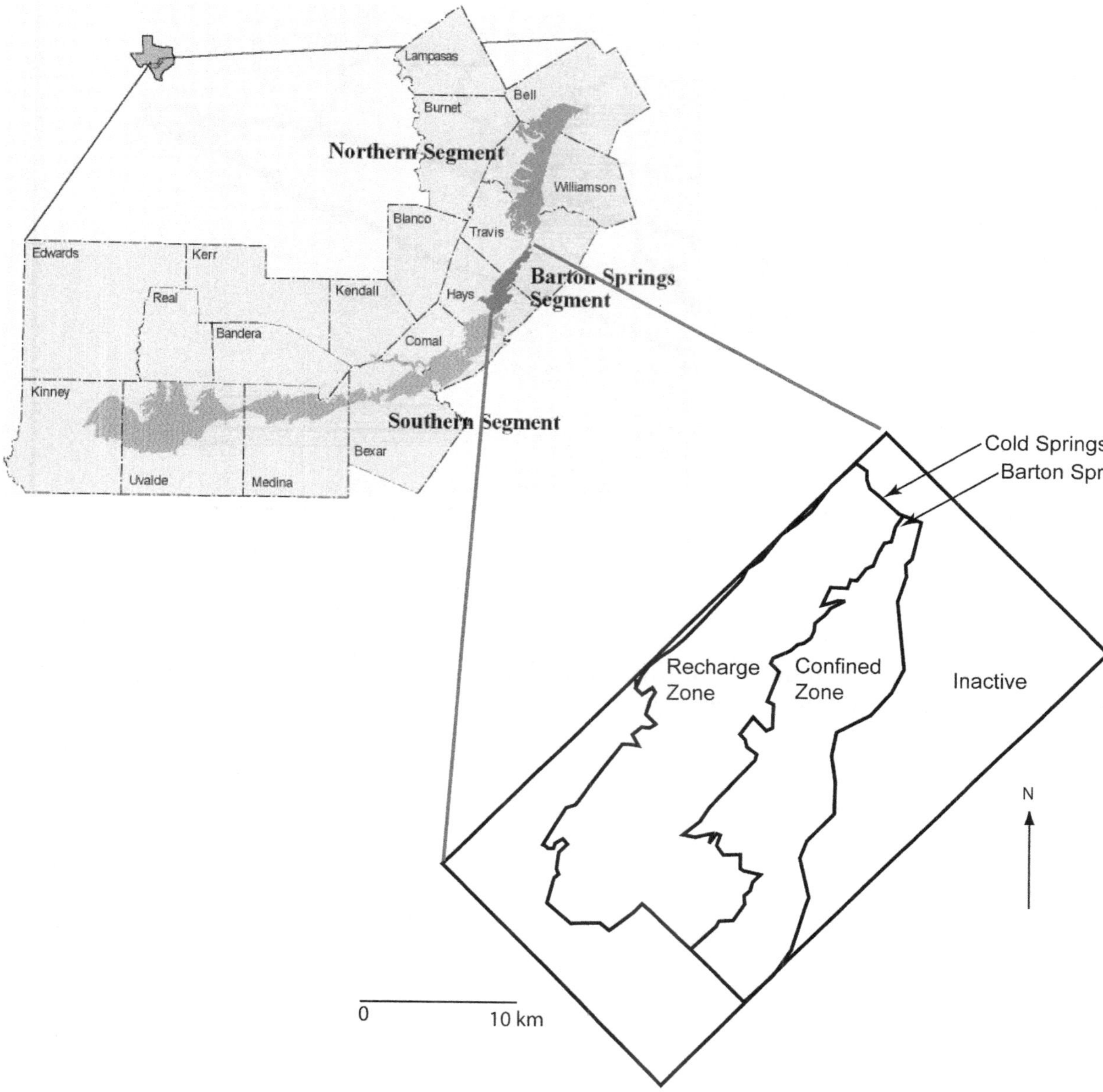

Figure 6.1 Study area for the Barton Springs model. The model domain is shown as an insert

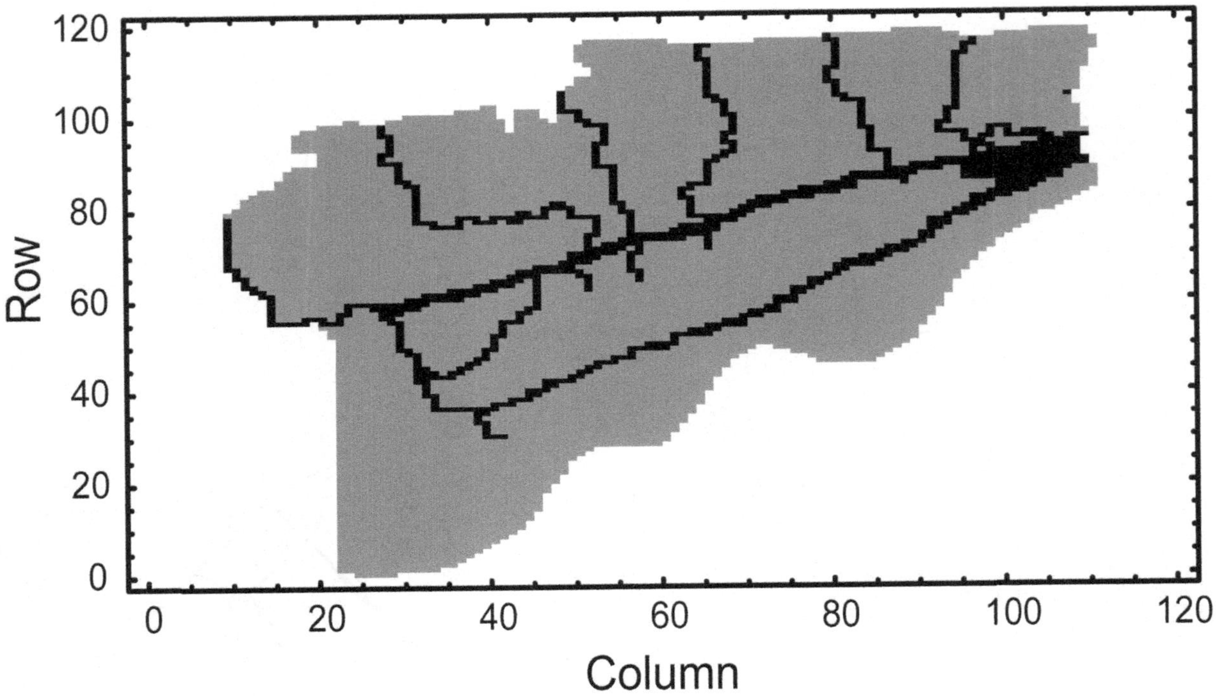

Figure 6.2 Locations of modeled conduits in the DCM model of Barton Springs. Each grid cell has dimensions 1,000 feet × 500 feet

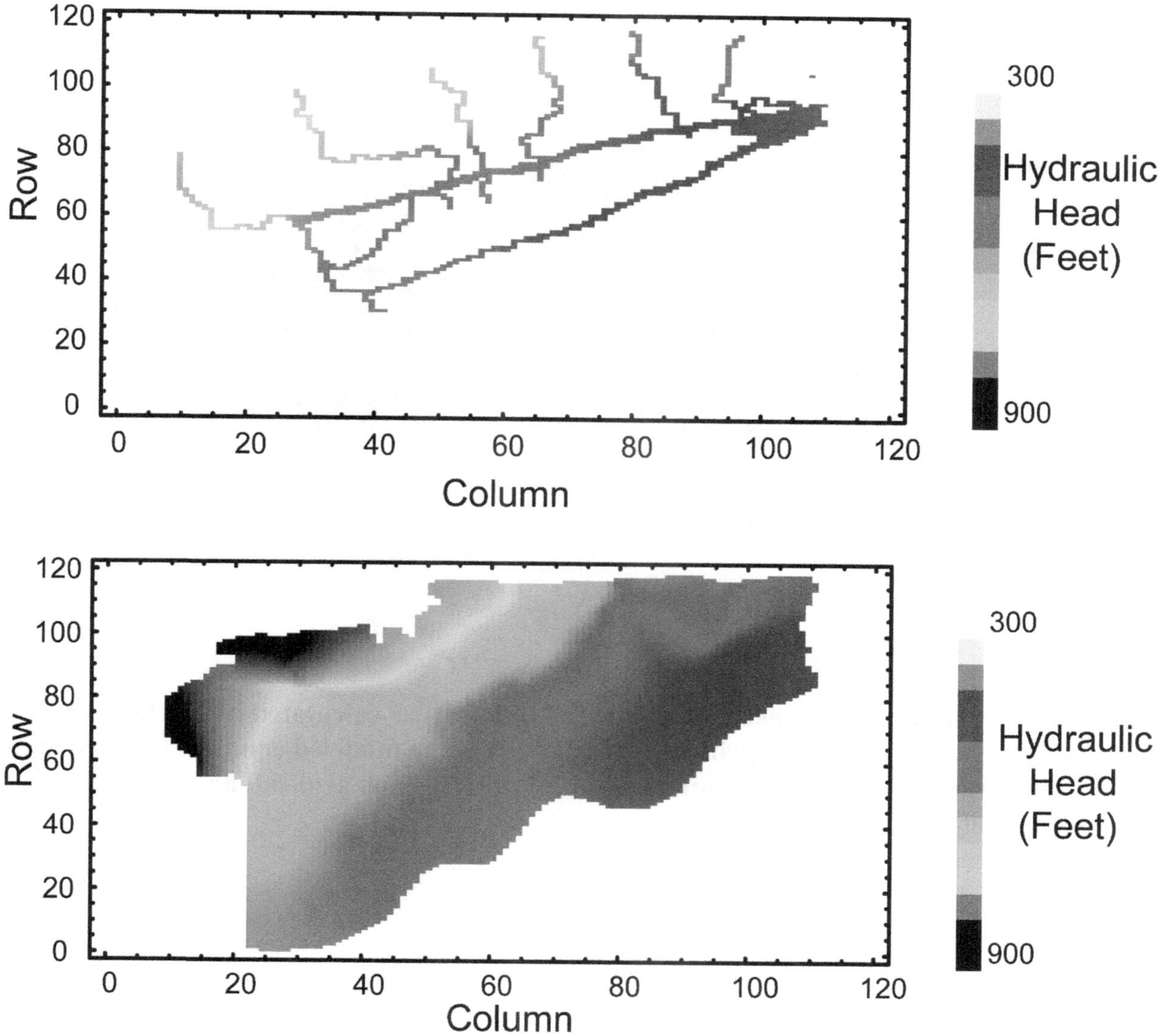

Figure 6.3 Calculated hydraulic heads in the conduit and diffuse systems for the Barton Springs steady-state model. Each grid cell has dimensions 1,000 feet × 500 feet.

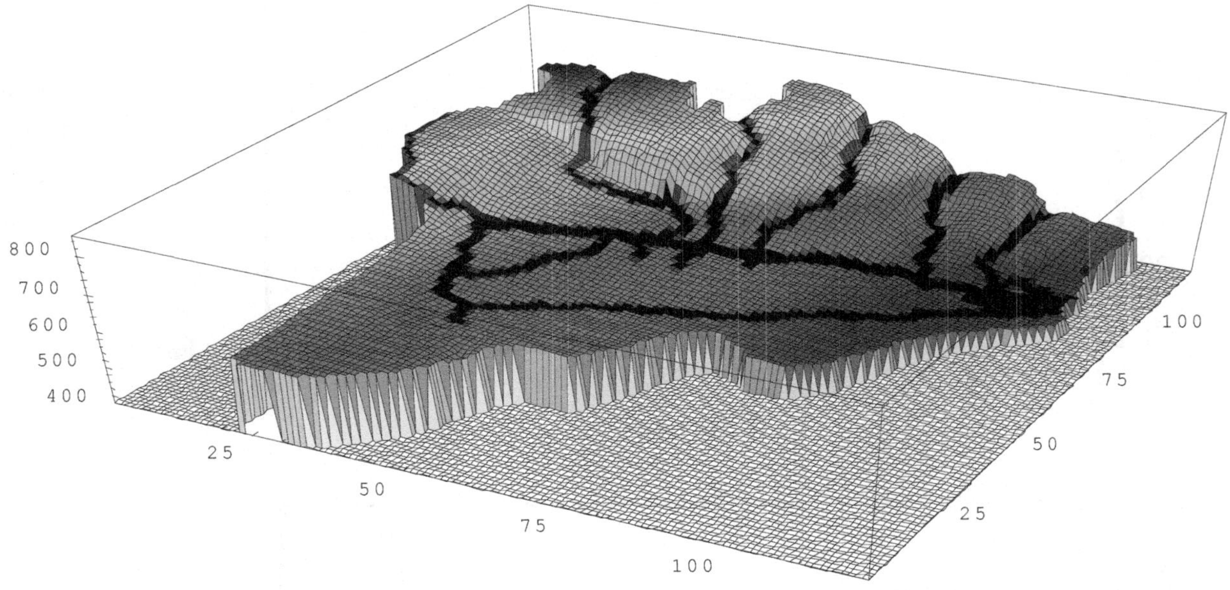

Figure 6.4 Surface plot of diffuse hydraulic heads (in feet) as calculated by the DCM steady-state model for Barton Springs. The locations of the modeled conduits are shown in black. The horizontal axes are column and row numbers. Each grid cell has dimensions 1,000 feet × 500 feet.

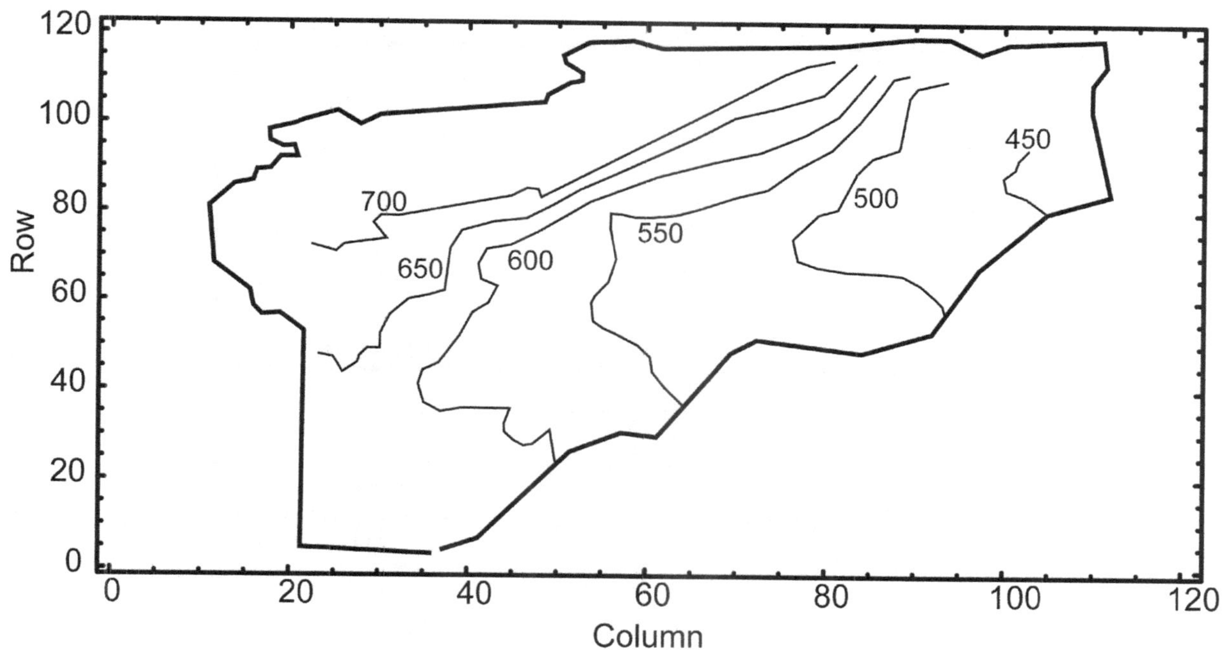

Figure 6.5 Contours of measured hydraulic heads (in feet) in the Barton Springs Segment of the Edwards Aquifer during the 1992 period (Scanlon et al., 2001, 2003). Each grid cell has dimensions 1,000 feet × 500 feet.

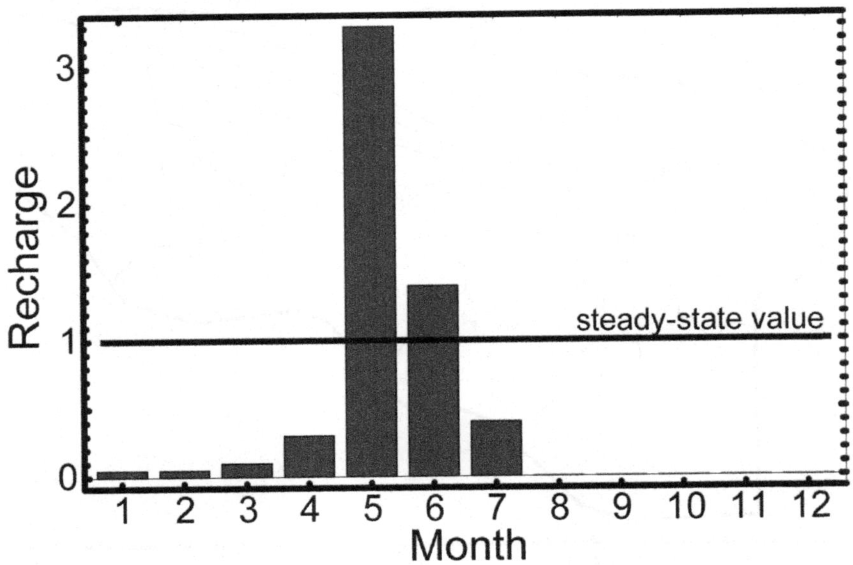

Figure 6.6 Monthly recharge multiplier used in the 1-year transient model of the Barton Springs Segment. Values shown are relative to the long-term average.

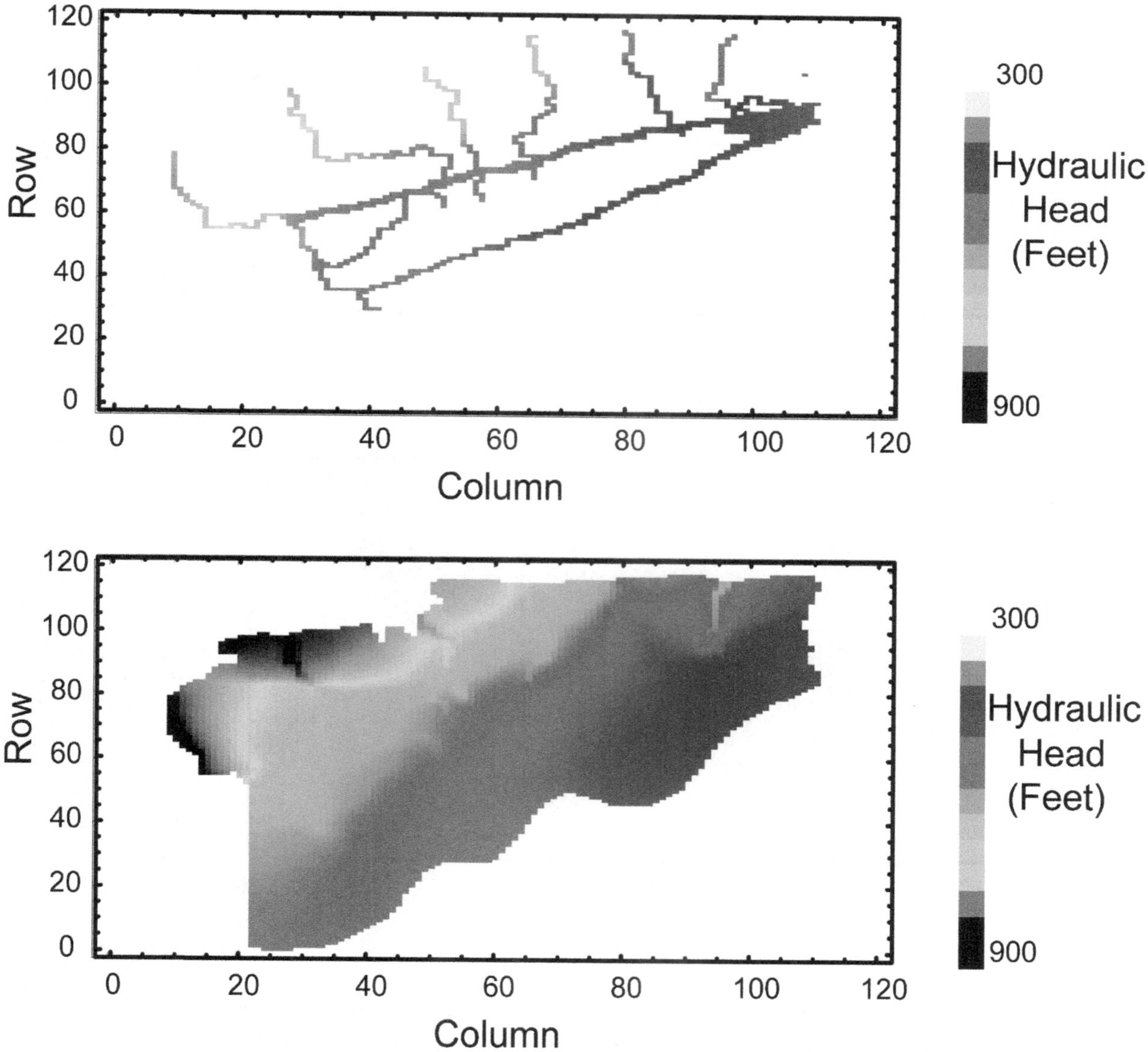

Figure 6.7 Calculated hydraulic heads in the conduit and diffuse systems at the end of month five in the Barton Springs 1-year transient simulation. Each grid cell has dimensions 1,000 feet × 500 feet.

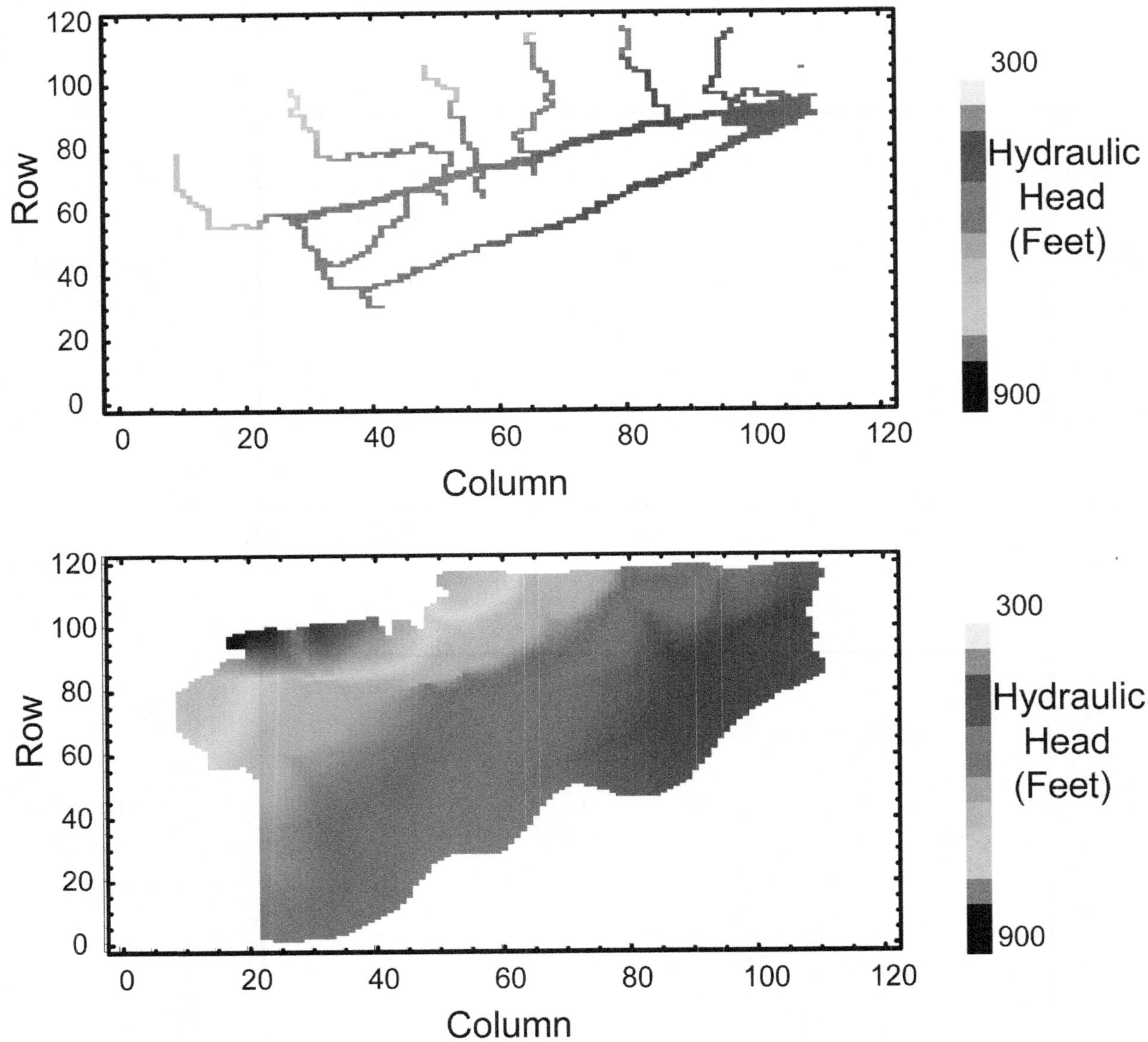

Figure 6.8 Calculated hydraulic heads in the conduit and diffuse systems at the end of the Barton Springs 1-year transient simulation. Each grid cell has dimensions 1,000 feet × 500 feet

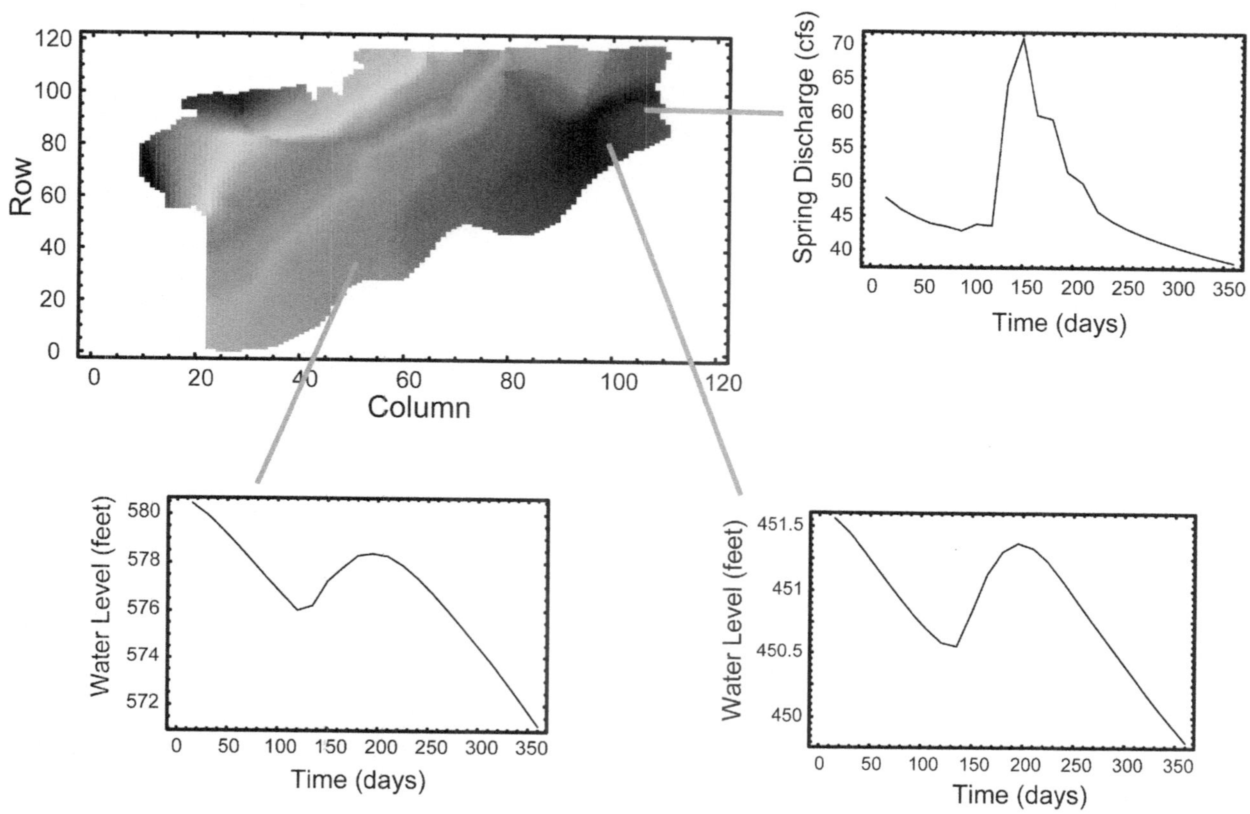

Figure 6.9 Spring-flow hydrograph and two water-level hydrographs from the Barton Springs transient simulation

CHAPTER 7
DISCUSSION AND CONCLUSIONS

MODELING APPROACHES FOR KARST AQUIFERS

Three approaches for modeling karst aquifers were evaluated in this project. The first approach evaluated was based on explicit representation of conduits embedded in a background diffuse flow system. The second approach used a smeared-conduit representation within existing single-continuum flow codes. The third approach used a dual-conductivity model. Other variants of these approaches are possible, but these three span the range of possible approaches for distributed parameter modeling of karst aquifers.

The embedded discrete-conduit model is intrinsically linked to the finite-element method. Although the approach is intuitively appealing, interaction between the conduit and the diffuse background—a potentially important calibration parameter—is difficult to control directly. Moreover, existing commercial finite-element codes appear to have inadequate implementation of discrete elements for unconfined conditions. For example, the commercial code FEFLOW®, which probably represents the most developed commercial finite-element flow code, does not include storage for partially filled conduits and does not allow transmissivity in a partially filled conduit to change with water level in the conduit. Finally, finite-element codes are not compatible with legacy data compiled primarily for the finite-difference code MODFLOW.

Numerical tests show the single-continuum, smeared-conduit approach—the traditional MODFLOW approach to modeling explicit conduits—can be calibrated to reproduce spring flow hydrographs. However, once calibrated to capture the rapid response of spring flow hydrographs to recharge events, the single-continuum approach tends to produce overly responsive water-level hydrographs. The magnitude of this error grows as the conduit density increases. In addition, the approach offers no explicit control of the diffuse-conduit interaction, which can be an important calibration parameter. In addition, MODFLOW does not have an option for a turbulent flow model, and modeling of unconfined karst aquifers in MODFLOW is questionable because of the well-known drying-rewetting instability.

Numerical experiments with the dual-conductivity model show that it offers increased flexibility compared with the single-continuum, smeared-conduit approach. In particular, the dual-conductivity model is able to simulate rapid spring responses without inducing overly responsive water-level hydrographs. The dual-conductivity approach introduces a new parameter describing the interaction between the conduit and diffuse system. This parameter must be determined by calibration.

THE DCM PACKAGE FOR MODFLOW-2000

The dual-conductivity approach was selected for further development, resulting in a new MODFLOW-2000 package for simulating karst aquifers within a dual-conductivity framework. Compatibility with legacy data was a key consideration in the decision to build this new modeling tool within the MODFLOW framework, and an important advantage in the selected approach. The new DCM Version 1.0 package was successfully validated by comparing to semi-analytical solutions and to alternative modeling software. Internal documentation and a user's manual were completed, consistent with Southwest Research Institute® Quality Assurance procedures.

Initial application of the DCM package to the Barton Springs Segment of the Edwards Aquifer demonstrates that the approach can be successfully used in practice. One-year transient simulations of the Barton Springs Segment produced flashy spring response, relatively subdued water level response, and transient groundwater troughs. These phenomena are all consistent with the observed behavior of the aquifer.

Although the initial experience with Barton Springs Segment suggests that explicit conduit models like DCM may improve the modeling of karst aquifers, it is important to recognize the range of appropriate use for the approach. Explicit conduit models require the modeler to specify locations for the conduits. In practice, conduit locations are highly uncertain. When modeling for the purposes of water supply management, the main quantities of interest are global water levels and spring flows. These time-dependent global quantities are sensitive to conduit properties and conduit connectivity, but are less sensitive to the exact locations of conduits. Thus, these global quantities may be estimated accurately even if the exact locations of conduits are not known. However, local water levels can be sensitive to the location of explicit conduits. For applications that require local water level information, independent field evidence to constrain conduit locations is needed. Explicit-conduit models should not be used to make deterministic predictions of local water levels in the absence of these independent constraints on conduit location. A more appropriate use of the technology in this situation would be to perform probabilistic assessment of water levels based on stochastic simulation of conduit location within an uncertainty band.

Data Requirements for Explicit Conduit Models

Specification of model parameters is a challenging aspect of any groundwater modeling study. Explicit conduit models require additional parameters compared with conventional porous media groundwater models. Understanding conduit data requirements and methods for estimating conduit parameters from field data is one of the long-term (future) goals of this project. Detailed case studies of multiple sites, planned for future phases of this project, will be required to fully develop this topic. However, some preliminary understanding has emerged based on the numerical experiments and the initial experience with the DCM package applied to the Barton Springs Segment.

Identifying the backbone of the conduit network is the most important requirement. Tracer-tests provide a well-established method for identifying connectivity between recharge features and springs, and are extremely valuable in developing a conceptual model for the conduit network. Major flow paths to springs corresponding to main conduits can often be identified from troughs in the potentiometric surface or from variations in water chemistry. The accuracy of this approach is sensitive to the density of monitoring wells in the vicinity of the conduit. If global quantities such as spring flows and large-scale averages of water levels are the only quantities of interest, then it may not be necessary to know the locations of conduits with high accuracy. Inferring conduit locations from groundwater troughs is likely to be easier during low-flow periods when the troughs are more pronounced. Obtaining water level measurements from several wells simultaneously (or nearly simultaneously) as part of a synoptic sampling program is also useful to minimize biases that can be introduced by changing water levels.

In unconfined systems, monitoring wells located near conduits may be able to provide information on the depth to partially filled conduits. For example, water levels in Well 58-50-411 in the Barton Springs Segment exhibits very minimal fluctuations (Scanlon et al., 2001), which can be attributed to the well's proximity to a large partially filled conduit at that elevation.[3]

Calibration is expected to be important for determining conduit properties such as conduit hydraulic conductivity and the conduit/diffuse exchange parameter. Specifically, conduit parameters can be adjusted either by trial-and-error or in a formal inverse procedure to force model spring flows to match measured spring flows. For calibration to be effective, accurate records of recharge and spring flows are necessary.

Additional Research Needs

Initial experience with the DCM package indicates that adaptive time-step control is needed to improve the practical utility of the modeling tool. In addition, DCM Version 1.0 has options for either turbulent or laminar flow in the conduits, but transition between laminar and turbulent flow in a single simulation is not supported. These refinements are planned for the second year of the project.

The other major activity for the second year of the project will be detailed demonstration simulations. The current plan is to continue with the Barton Springs Segment. In addition, a segment of the Floridan aquifer will be identified and modeled. These two field sites are especially attractive for detailed demonstration simulations because many of the karst characteristics that are key to the successful implementation of MODFLOW-DCM have been extensively investigated. In particular, these characteristics include accurate tracer mapping of conduits, precise hydrographs (both in the matrix and in spring outflows), high resolution recharge (i.e., precipitation), and measurement of matrix and conduit hydraulics. Although not all of these parameters have been characterized to the degree desired for simulation, it is recognized that these two sites offer data support that is more substantial than most other potential demonstration sites. And although it is also recognized that most karst aquifers that require numerical models for proper groundwater resource management will not have this level of data support, it is imperative to first demonstrate the utility and limitations of MODFLOW-DCM at sites that are reasonably well characterized.

The longer term research needs for improving karst modeling with DCM are to link DCM with a contaminant transport modeling tool, adapt formal automatic calibration methods (inverse modeling) to work with DCM, develop a 3D version of DCM, and develop stochastic simulation methods to address uncertainty in conduit positions.

Although not specifically address in this project, improved geophysical methods to help locate conduits would be a valuable complement to the new modeling tools.

[3] N.M. Hauwert, 2004, personal communication

REFERENCES

Anderman, E.R., K.L. Kipp, M.C. Hill, J. Valstar, and R.M. Neupauer. 2002. MODFLOW-2000, the U.S. Geological Survey Modular Ground-Water Model—Documentation of the Model-Layer Variable-Direction Horizontal Anisotropy (LVDA) Capability of the Hydrogeologic-Unit Flow (HUF) Package. U.S. Geological Survey Open-File Report 02-409. Denver, Colo.

Atkinson, T.C. 1977. Carbon Dioxide in the Atmosphere of the Unsaturated Zone: An Important Control of Groundwater Hardness in Limestones. *Journal of Hydrology*, 35:111–123.

Barenblatt, G.I., I.P. Zheltov, and I.N. Kochina. 1960. Basic Concepts in the Theory of Seepage of Homogeneous Liquids in Fissured Rocks. *Journal of Applied Mathematics and Mechanics*, 24:1,286–1,303.

Budd, D.A., and H.L. Vacher. 2004. Matrix Permeability of the Upper Floridan Aquifer. *Hydrogeology, forthcoming.*

Calic-Ljubojevic, J. 2002. Bedding-Plane Anastomoses as one of the Early Stages of Karst Evolution. *Acta Geologica Polonica*, 52:111–115.

Clemens, T., D. Hückinghaus, M. Sauter, R. Liedl, and G. Teutsch. 1996. A Combined Continuum and Discrete Network Reactive Transport Model for the Simulation of Karst Development. In *Calibration and Reliability in Groundwater Modelling*. Edited by K. Kovar, and P. van der Heijde. Golden, Colo. IAHS Publ.

Diersch, H.-J. G. 2002. FEFLOW—Reference Manual. Berlin: WASY Ltd.,

Faulkner, G.L. 1976. Flow Analysis of Karst Systems With Well Developed Underground Circulation, Karst Hydrology and Water Resources. In *Proceedings of United States–Yugoslavian Symposium, Dubrovnik, Croatia. June 2–7, 1975*: Water Research Publ. p. 6-1 to 6-28.

Gale, S.J. 1984. The Hydraulics of Conduit Flow in Carbonate Aquifers. *Journal of Hydrology*, 70:309–327.

Halihan, T., C.M. Wicks, and J.F. Engeln. 1998. Physical Response of a Karst Drainage Basin to Flood Pulses: Example of Devil's Icebox Cave System (Missouri, USA). *Journal of Hydrology*, 204:24–36.

Halihan, T., J.M. Sharp, and R.E. Mace. 2000. Flow in the San Antonio Segment of the Edwards Aquifer: Matrix, Fractures, or Conduits? In *Groundwater Flow and Contaminant Transport in Carbonate Aquifers*. Edited by I.D. Sasowsky and C.M. Wicks. Rotterdam: Balkema.

Hauwert, N.M, D.A. Johns, J.W. Sansom, and T.J. Aley. 2002. Groundwater Tracing of the Barton Springs Edwards Aquifer, Travis and Hays Counties, Texas. *Gulf Coast Association of Geological Societies. Transactions*, 52: 377-384

Harbaugh, A.W., E.R. Banta, M.C. Hill, and M.G. McDonald. 2000. MODFLOW-2000, the U.S. Geological Survey Modular Ground-Water Model—User Guide to Modularization Concepts and the Ground-Water Flow Process. U.S. Geological Survey Open-File Report 00-92. Denver, Colo.

Hess, J.W., and W.B. White. 1988. Storm Response of the Karstic Carbonate Aquifer of Southcentral Kentucky. *Journal of Hydrology*, 99:235–252.

Hill, C.A. 2000. Overview of the Geologic History of Cave Development in the Guadalupe Mountains, New Mexico. *Journal of Cave and Karst Studies*, 62(2):60–71.

Jeannin, P.-Y. 2001. Modeling Flow in the Phreatic and Epiphreatic Karst Conduits in the Holloch Cave (Muotatal, Switzerland). *Water Resources Research*, 37:191–200.

Kaufmann, G. and J. Braun. 1999. Karst Aquifer Evolution in Fractured Rocks. *Water Resources Research*, 35(11):3,223–3,238.

Kiraly, L. 1998. Modelling Karst Aquifers by the Combined Discrete Channel and Continuum Approach. *Bulletin d'Hydrogeologie*, 16:77–98.

Kiraly, L. 2003a. Karstification and Groundwater Flow. In *Speleogenesis and Evolution of Karst Aquifers*, 1(3).

Kiraly, L. 2003b. Karstification and Groundwater Flow. In *Evolution of Karst: From Prekarst to Cessation*. Edited by F. Gabrovsek. Postojna-Ljubljana, Zalozba: ZRC. Pp. 155–190.

Klimchouk, A.B. and D.C. Ford. 2000a. Types of Karst Evolution of Hydrogeologic Setting. In *Speleogenesis: Evolution of Karst Aquifers*. Edited by A.B. Klimchouk, D.C. Ford, A.N. Palmer, and W. Dreybrodt. Huntsville, AL: National Speleological Society, Inc. pp. 45–53.

Klimchouk, A.B., and D.C. Ford. 2000b. Lithologic and Structural Controls of Dissolutional Cave Development. In *Speleogenesis: Evolution of Karst Aquifers*. Edited by A.B. Klimchouk, D.C. Ford, A.N. Palmer, and W. Dreybrodt. Huntsville, AL: National Speleological Society, Inc. pp. 54–64.

Klimchouk, A.B. 2003. Unconfined Versus Confined Speleogenetic Settings: Variations of Solution Porosity and Permeability. Proceedings of the International Conference on Karst, *Hydrogeology and Ecosystems,* June 3–6, 2003. Bowling Green, Kentucky.

Leake, S.A., and D.V. Claar. 1999. Procedures and Computer Programs for Telescopic Mesh Refinement Using MODFLOW. U.S. Geological Survey Open-File Report 99-238. Denver, Colo: USGS.

Murdoch, J.W. 1996. Mechanics of Fluids. In *Marks' Standard Handbook for Mechanical Engineers, 10th Edition*. Edited by E.A. Avallone, and T. Baumeister III. New York: McGraw Hill.

Mylroie, J. 2003. The Interaction of Hypogenic Caves and Explorational Bias: Under Representation of Cave Data. In Proceedings of the International Conference on Karst Hydrogeology and Ecosystems, June 3–6, 2003. Bowling Green, Kentucky.

Palmer, A.N. 1981. Hydrochemical Controls in the Origin of Limestone Caves. Paper presented at the 8th International Speleological Congress, Bowling Green, Ky. National Speleological Society, Inc.

Palmer, A.N. 1991. Origin and Morphology of Limestone Caves. *Geological Society of America Bulletin*, 103:1–21.

Scanlon, B.R., R.E. Mace, B. Smith, S. Hovorka, A.R. Dutton, and R. Reedy, 2001. *Groundwater Availability Modeling of the Barton Springs Segment of the Edwards Aquifer, Texas: Numerical Simulations Through 2050.* Austin, TX. Bureau of Economic Geology.

Scanlon, B.R., R.E. Mace, M.E. Barrett, and B. Smith. 2003. Can We Simulate Regional Groundwater Flow in a Karst System Using Equivalent Porous Media Models? Case study, Barton Springs Edwards Aquifer, USA. *Journal of Hydrology*, 276:137–158.

Shuster, E.T., and W.B. White. 1971. Seasonal Fluctuations in the Chemistry of Limestone Springs: A possible Means for Characterizing Carbonate Aquifers. *Journal of Hydrology*, 14:93–128.

Springer, G.S. 2004. A Pipe-Based, First Approach to Modeling Closed Conduit Flow in Caves. *Journal of Hydrology*, forthcoming.

Svensson, U. 2001. A Continuum Representation of Fracture Networks, Part I: Method and Basic Test Cases. *Journal of Hydrology*, 250:170–186.

Teutsch, G., and M. Sauter. 1998. Distributed Parameter Modeling Approaches in Karst-Hydrological Investigations. *Bulletin d'Hydrogeologie,* pp. 16:99–109.

White, W.B. and E.L. White. 2003. Conduit Fragmentation, Cave Patterns, and the Localization of Karst Ground Water Basins: The Appalachians as a Test Case. *Speleogenesis and Evolution of Karst Aquifers*, 1(2).

Worthington, S.R.H., D.C. Ford, and P.A. Beddows. 2000. Porosity and Permeability Enhancement in Unconfined Carbonate Aquifers as a Result of Solution. In *Speleogenesis: Evolution of Karst Aquifers.* Edited by A. Klimchouk, D.C. Ford, A.N. Palmer, and W. Dreybrodt. Huntsville, Alabama: National Speleological Society, Inc. pp. 463–472.

Worthington, S.R.H. 2003. A Comprehensive Strategy for Understanding Flow in Carbonate Aquifer. *Speleogenesis and Evolution of Karst Aquifers*, 1(1).

ABREVIATIONS

AwwaRF	Awwa Research Foundation
BSEACD	Barton Springs/Edwards Aquifer Conservation District
CAVE	Carbonate Aquifer Void Evolution
cm	centimeter
CNWRA	Center for Nuclear Waste Regulatory Analyses
DCM	dual-conductivity model
EAA	Edwards Aquifer Authority
FEFLOW	Finite Element Subsurface Flow and Transport Simulation System
km	kilometer
LPF	Layer-Property-Flow
LVDA	Layer Variable-Direction horizontal Anisotropy
m	meters
m^{-1}	per meter
m/s	meter per second
m^2	square meters
m^2/min	square meters per second
m^3/s	cubic meters per second
m^3/day	cubic meters per day
Min	minute
MODFLOW	MODular groundwater FLOW
s	seconds
SWFWMD	Southwest Florida Water Management District
t	time
1D	one-dimensional
2D	two-dimensional
3D	three-dimensional

IWA Publishing's authorised EU representative for General Product Safety Regulations is Diane D'Arras, 15 rue Duret, 75116 Paris, France, e-mail: safety@iwap.co.uk.

Printed and bound by CPI Group (UK) Ltd, Croydon, CR0 4YY
17/09/2026
02218810-0012